Lecture Notes in Statistics

Edited by D. Brillinger, S. Fienberg, J. Gani,
J. Hartigan, and K. Krickeberg

14

GLIM 82: Proceedings of the International Conference on Generalised Linear Models

Edited by Robert Gilchrist

Springer-Verlag
New York   Heidelberg   Berlin

Robert Gilchrist
The Polytechnic of North London
Department of Mathematics
Holloway, London N7 8DB
England

AMS Classification: 6204

Library of Congress Cataloging in Publication Data
International Conference on Generalised Linear
    Models (1st : 1982 : Polytechnic of North London)
    GLIM 82 : proceedings of the International Conference
on Generalised Linear Models.

    (Lecture notes in statistics ; 14)
    Conference held Sept. 13-15, 1982.
    1. Linear models (Statistics)--Congresses. I. Gil-
christ, Robert. II. Title. III. Title: GLIM eighty two.
IV. Title: G.L.I.M. 82. V. Series: Lecture notes in
statistics (Springer-Verlag) ; 14.
QA276.A1I537 1982   519.5   82-16728

With 52 Illustrations

9 8 7 6 5 4 3 2 1

ISBN-13: 978-0-387-90777-2   e-ISBN-13: 978-1-4612-5771-4
DOI: 10.1007/978-1-4612-5771-4

PREFACE

This volume of Lecture Notes in Statistics consists of the published proceedings
of the first international conference to be held on the topic of generalised linear
models. This conference was held from 13 - 15 September 1982 at the Polytechnic of
North London and marked an important stage in the development and expansion of the
GLIM system. The range of the new system, tentatively named Prism, is here outlined
by Bob Baker. Further sections of the volume are devoted to more detailed descriptions
of the new facilities, including information on the two different numerical methods
now available. Most of the data analyses in this volume are carried out using the
GLIM system but this is, of course, not necessary. There are other ways of analysing
generalised linear models and Peter Green here discusses the many attractive features
of APL, including its ability to analyse generalised linear models.
Later sections of the volume cover other invited and contributed papers on the theory
and application of generalised linear models. Included amongst these is a paper by
Murray Aitkin, proposing a unified approach to statistical modelling through direct
likelihood inference, and a paper by Daryl Pregibon showing how GLIM can be
programmed to carry out score tests.
A paper by Joe Whittaker extends the recent discussion of the relationship between
conditional independence and log-linear models and John Hinde considers the
introduction of an independent random variable into a linear model to allow for
unexplained variation in Poisson data.

Other papers cover a fairly wide range of topics. Amongst the areas of application
are genetics, insecticide evaluation trials and cot-death data. More theoretical
papers consider binary response models with misclassification, parametric link
functions and multiple comparisons.
The conference was organised under the auspices of the Royal Statistical Society
Working Party on Statistical Computing; the programme committee was M.Aitkin, R.J.Baker,
M.R.B.Clarke, R.Gilchrist, M.Green, J.A.Nelder, C.Payne and M.Richardson. Contributed
papers were refereed by the committee with additional help from K.V.C.Kennard.

RG

Department of Mathematics,
Statistics and Computing,
Polytechnic of North London, UK.

CONTENTS

A GLM FOR ESTIMATING PROBABILITIES IN RETROSPECTIVE CASE-CONTROL STUDIES

M.Slater, Queen Mary College, U.K. and R.D. Wiggins,
Polytechnic of Central London, U.K.

INTRODUCTION

J.A. Nelder

Rothamsted Experimental Station

Harpenden, Herts., AL5 2JQ, U.K.

It is now ten years since the late Robert Wedderburn and I published our paper defining generalised linear models, and the papers presented in this volume give a good indication of how the idea has been taken up and extended since then.

The development of GLIM, a program for fitting generalised linear models and designed to be used interactively, has undoubtedly helped to propogate the ideas underlying such models. The use of the syntax developed by Wilkinson and Rogers for specifying complex linear models made the description of a generalised linear model to the computer simple, it being only necessary to supply in addition the probability distribution involved and the link function. GLIM3 is now in use in more than 500 centres throughout the world, and besides its basic use for analysing data has been used quite widely as a teaching aid. Its uses have extended well beyond the originally intended class of models. In particular various forms of analysis of survival data have been found to be expressible in GLIM terms. The introduction of quasi-likelihood by Wedderburn has allowed the replacement of the full distributional assumption by assumptions about the variance-mean relation only, thus greatly enlarging the class of models.

The new system will contain the GLIM algorithm, with extended facilities, as one of its modules. Another module will deal with computer graphics, and provide the means of constructing pictures by which the fit of a model can be inspected by that unrivalled detector of patterns, the human mind and eye. Pioneering work by Pregibon on model checking for generalised linear models will allow the ideas developed for classical linear models to be extended to the wider class.

The papers that follow will describe the various modules of the new system, the underlying algorithms used for fitting generalised linear models, and a variety of applications. From these the reader will obtain a good idea of the scope of these models in statistics.

The new system will include, as a further module, the algorithm for the analysis of experiments with generally balanced designs, originally developed for Genstat by Wilkinson, Rogers and Payne. This very general algorithm, which uses two model formulae of the GLIM type to describe the structure of the design, allows, for example, the simple specification and analysis of all designs in Cochran and Cox's standard text. Data can be moved easily to and from the other modules in the new system.

The new system will thus offer the statistician an integrated set of tools for fitting a large class of statistical models and for inspecting the results of such a fitting. By doing so it should encourage the intelligent scrutiny of both data and models that good statistical practice requires.

PRISM - AN OVERVIEW

R.J. BAKER

Rothamsted Experimental Station
Harpenden
Herts, U.K.

SUMMARY

The statistical program Prism is introduced as a development of GLIM-3. The features of GLIM-3 that made it simple and convenient for analysing generalised linear models have been consolidated in the new program. Some new language features of Prism are outlined and the data structures now supported by the System are described. Finally a brief description of the data manipulation and program control facilities of Prism is given.

Keywords: GLIM, GENERALISED LINEAR MODELS, PROGRAMMING LANGUAGES, GRAPHICS, ARRAYS, ANALYSIS-OF-VARIANCE, DATA STRUCTURES, STATISTICAL PACKAGES, TABLES

1. INTRODUCTION

Prism is a development from GLIM-3. The latter was released in 1978 and has since become very popular, mainly for the ease with which it enables generalised linear models to be analysed. But users also appreciated the simplicity and flexibility of its command language and Prism has been developed with the twin aims of both enriching and consolidating that command language and extending the areas of statistical analysis that can be handled.

We shall see shortly that Prism is composed of <u>four</u> 'modules' - the Kernel (for data manipulation and program control), the GLIM (Release 4) module, the AOV (analysis-of-variance) module and the GRAPH (graphics) module. This paper gives an overview of Prism but will concentrate on the features of the Kernel. Papers by Clarke (1982), Payne (1982) and Slater (1982) at this Conference describe in more detail the GLIM, AOV and GRAPH modules respectively. Additionally, Green (1982) discusses, at greater length than here, the facilities in the Kernel for manipulating multidimensional arrays and tables.

(It is possible that some of the features described later in this paper will not be available in Release 1 of Prism. In particular the full HELP and FEEDBACK facilities and the SQUEEZE directive may not become available till Release 2. Similarly, minor changes to syntax may occur between the writing of this paper and the release of Prism. In all cases the Prism Reference Guide will take precedence over any definitions or declarations given here).

In the next section we give some of the historical background to the development of Prism, noting those features of GLIM-3 that Prism was intended to build on, and in section 3 we discuss how these have been implemented. Section 4 discusses some of the general features of the Prism language. Sections 5 and 6 describe the data structures supported by Prism, section 5 concentrating on the implementation of multidimensional arrays and tables, while sections 7 and 8 describe respectively the data manipulation and the program control facilities provided in the Kernel.

2. BACKGROUND

GLIM-3 was first distributed in 1978, see Baker and Nelder (1978). Its success has been quite remarkable and is I believe attributable to several factors.

2.1. *Analysing glms*

First and foremost, GLIM-3 enabled generalised linear models to be analysed easily and flexibly. The pioneering paper of Nelder and Wedderburn (1972) pointed out the theoretical framework common to many of the univariate models used in statistics and showed that a unified approach to their analysis was possible. This theoretical framework was later reflected in the structure of the GLIM commands, i.e. the ERROR, LINK and FIT directives. Additionally, the structure formulae devised by Wilkinson and Rogers (1973) enabled complex explanatory relationships to be specified through direct combinations of factors and variates, the somewhat novel treatment of redundancy among the linear parameters freed the user from the computational complications of aliasing, and the ability to fit any sequence of individual models contrasted with the 'all-at-once' approach of most analysis-of-variance programs, but made it possible to fit models to unbalanced data in whichever order was appropriate.

2.2 *Interactive Use*

A second factor contributing to the popularity of GLIM was its interactive mode of use. In the early seventies there were few generally available statistical programs that could be used interactively, most statistical analysis being carried out in batch. Even then, and certainly more so now, this was recognised as undesirable for many types of analysis, particularly those of an exploratory nature, where immediate interaction with the program is needed.

2.3. *Simple Command Language*

Thirdly, although very complex analyses could be performed by the program, the basic command language through which this was achieved remained relatively straightforward. For example, to specify a prior weight W, only the statement $WEIGHT W was needed, while the statement needed to remove it was even simpler. Though novice users may have had difficulties with some of the statistical concepts involved in GLIM analyses, most found the command language and its syntax both easy to learn and simple to use.

2.4. *Program Control*

Finally, and related to the previous point, the command language, though elementary, could be extended to arbitrary complexity through the provision of macro structures and their use in the USE, WHILE, SWITCH etc. directives. Sophisticated use of macros enabled sets of instructions for simple data manipulation to be stored for routine use in the future, as exemplified by the Hull macro library, and, in the case of the macros produced at the University of Lancaster, their use extended the class of statistical models that could be analysed beyond the limits originally envisaged by the developers of the program.

2.5. *The Beginnings of Prism*

Following this successful development of GLIM-3 there was a demand for an interactive analysis-of-variance package based on the same algorithm as the ANOVA directive in Genstat (see Alvey et al, 1977) but set into the syntax of GLIM. Similarly the current, much-needed emphasis on graphical methods in statistical analysis led to the investigation of ways to incorporate graphical capabilities into GLIM. Now, it would have been possible to develop each project separately, resulting in three distinct packages. Or all three projects could have been combined in one large package. But the first possibility has a drawback. Since most operating systems still do not provide satisfactory means of communicating between programs, the user wishing to pass a subset of GLIM data to the analysis-of-variance package for its specialized analysis, or the user wishing to iterate between, say, fitting and plotting residuals, would face great difficulties. On the other hand many computers for which GLIM is designed have storage restrictions that would prevent the loading of the large, combined package.

2.6. *Its Modular Structure*

The solution chosen was to develop the projects as separate <u>but</u> compatible modules. Release 4 of the generalised linear interactive modelling package became the GLIM module, the analysis-of-variance module was called AOV and the graphics handler was called GRAPH. In order to avoid duplication of code etc., a 'Kernel' was defined that contained those facilities that would be needed by all three projects, resulting in a total of four modules. The Kernel plus any (including the null) subset of the 3 'external' modules can be used to form an executable program. For example the Kernel + GLIM constitutes the successor to GLIM-3, while the Kernel + AOV + GRAPH is an analysis-of-variance program with full graphical capabilities. The complete System has provisionally been entitled Programs for interactive statistical modelling, or Prism for short. The arrangement is shown diagrammatically below

Prism

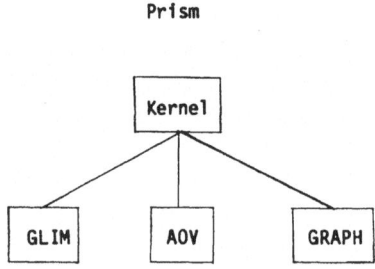

Sites with space problems may wish to mount only Kernel + GLIM to keep the program as small as possible. Additionally, those sites needing a high-level data manipulation and graphical display program may find Kernel + GRAPH precisely what they require. The modular approach offers the site flexibility in the way it implements Prism.

An intrinsic advantage of the above approach is that a unified command language is possible across the modules, so that users have only one system and one command language to master. This command language is an exension of that used in GLIM-3 and in the next section we discuss how this and many of the other features of GLIM have carried over to Prism.

3. SOME GENERAL CONSIDERATIONS

The underlying approach to the development of Prism has been to preserve and consolidate those features mentioned in the previous section that GLIM-3 users found most attrctive while improving those areas (such as the output from the program) that gave rise to most complaints.

3.1. *Analysing glms*

From the point of view of analysing glms the GLIM module has both simplified the fitting of glms and widened the range of models that can be fitted. Firstly, the introduction of optionwords and the rationalization of structure formulae have simplified the specification and fitting of glms, while the enhanced DISPLAY and EXTRACT directives and the introduction of the PREDICT directive have greatly eased the production of fitted and predicted values and other output from a fit. Secondly the user now has both a wider choice of model (new probability distributions and link functions including composite links) and greater control over the numerical method including the ability to trade off the space used against time taken to perform a fit. See Clarke (1982) for more details.

Additionally the AOV module can be used to perform the special analysis-of-variance computations that are possible for data from generally balanced designs, see Payne (1982), while the GRAPH module provides a general-purpose set of tools for generating, manipulating and displaying graphical data in an interactive environment. Both these modules considerably enhance the facilities of the GLIM module in analysing glms.

3.2. *Interactive use*

It was, of course, axiomatic that Prism, like GLIM, would be interactive - the second feature of GLIM-3 mentioned in section 1. But it is now the GRAPH module that has benefitted most from this feature: the ability to interactively input, manipulate and redisplay graphical data on a modern raster screen is as different from the batch graphics of plotters and storage tubes as the early GLIM was from the statistical packages of the sixties. Perhaps GRAPH is the graphical GLIM of the eighties!

3.3. *The Command Language*

The Prism command language differs from that of GLIM-3 in that it contains several new language features additional to those of GLIM-3. (These features are described in section 4.) However the Prism language is not a superset of the GLIM-3 language, as certain GLIM-3 language features are inconsistent with those of Prism or were considered undesirable. For example the 'pathwords' of GLIM-3 (i.e. the names such as NORMAL, IDENTITY, E, of the ERROR, LINK, DISPLAY, etc directives) were allowed to be of any length in the ERROR directive but of a single character in the DISPLAY directive where concatenated lists were allowed. In Prism, however, all pathwords are significant up to 4 characters but may be of any length; this is obviously inconsistent with GLIM-3 usage so that although the statement

$display   ERC

was valid under GLIM-3, it will be faulted under Prism, ERC not being a valid DISPLAY pathword. Thus care must be taken when running GLIM-3 programs under Prism. An appendix of the Reference Guide will list the changes between GLIM-3 and Prism that are likely to cause problems.

On the positive side it has been possible to define a language that not only retains the simplicity of GLIM-3 but also, through the new language features, is easier to use. Besides the features described in section 4 the following additions in Prism are worth mentioning: missing values (input as *) are accepted almost everywhere and the use of a * has a well-defined meaning wherever it is accepted; similarly a * may be used in place of a pathword, in which case the default action is taken; lower case characers are always acceptable to the program and are interchangeable with their upper-case equivalents, so that 'ABcd' and 'abCD' denote the same identifier; the underline symbol _ may appear embedded in an identifier; in many places where an identifier must be supplied to hold output values (e.g. the EXTRACT directive) an 'io-request' symbol <> may be supplied instead, in which case the values will be printed at the terminal; similarly if the io-request symbol is used when input values are required (e.g. in the PLOT directive) these will be taken from the terminal.

3.4. *Program Control*

Prism now provides an extensive program-control capabilitiy, representing a considerable improvement over that of GLIM-3. (A short description is given in section 8.) Since approximately half of Prism's language interpreter is concerned with program control it is worth justifying the effort devoted to this aspect of the System.

An obvious advantage is that it becomes possible to store a set of routine instructions for an analysis as a macro, making the analysis more convenient to perform. A more important point is that not all the facilities that users, or even the program developers, would wish for can be included in a given package. Many of these will not be forseen by the developers, and between those that can be forseen a choice will be made on the grounds of size and cost of coding and usefulness to users. Consequently the user of a statistical package will always find that the package does not perform all the tasks he/she requires, or performs them in an inconvenient manner for his/her purposes. The provision of program-control capabilities is an essential feature of any statistical package, because it enables the user to extend or rearrange the basic capabilities of the package.

3.5. *Output From Prism*

The commonest complaints about GLIM-3 concerned its output layout: the formatting of numbers was wrong (not enough or too many significant figures), it was too terse, it was very difficult to produce tables of output (or predicted) values and, trivial but annoying, the value zero was printed in E-format. In Prism we have attempted to improve the program's output wherever possible.

The principle reason for the poor formatting of GLIM-3 output lay in the decision to work through the Fortran FORMAT statement, and then only through its portable version. Designed for use when relatively simple output is produced line-by-line and when a fixed format is required, the FORMAT statement is otherwise inflexible and restricting. (In particular the appearance of the value zero under G-format is out of the control of the programmer.) Hence, it was decided to write a suite of subroutines to perform flexible, item-by-item (as in Pascal) output in a modular, portable way. (The routines are also available as a general software tool - see Henstridge et al, 1982).

It is now possible to format numbers effectively, choosing the number of digits and the number of decimal places appropriate to the whole set of numbers to be displayed. Similarly it is possible to give the user some control over the number of significant figures printed: for each directive a default number of significant figures is defined as a standard accuracy for that particular output, but this may be increased by the user through the ACCURACY option of the OUTPUT directive. For example, the LOOK directive usually prints to 4 significant figures, but this may be increased to 6 by using

$output   ACCURACY=2

('ACCURACY=' is an optionword to be described later.)

The OUTPUT directive also contains an option to print or suppress warnings, as well as the usual settings for the output channel number, width and height. These settings have a global effect. If, however, a setting is required for the duration of a single directive then this may be achieved through a 'temporary output setting', which has the same syntax as the OUTPUT directive but is enclosed in angle brackets and follows immediately the directive-name. Thus in the above example a temporary setting for the LOOK directive could be achieved by:

$look< ACCURACY=2 >

The output from the GLIM module has also been considerably enhanced, more information being provided by the FIT directive, the DISPLAY output being better formatted, the EXTRACT directive permitting values to be printed directly and the PREDICT directive enabling the simple production of tables of fitted values, etc (see Clarke, 1982, for more details). In the Kernel the ENVIRONMENT directive gives clearer and more detailed information; similarly the OBTAIN directive in AOV and the INQUIRE directive in GRAPH allow the user to extract information about the state of the module, perhaps into the user's structures so that the information can be stored, or else for display at the terminal.

Care has been taken to make fault messages more positive ('a factor is needed here' rather than 'invalid identifier type') and to provide more, but optional, information on its location.

The only area in which there have been few concessions concerns the amount of output produced. We are aware that for interactive use the production of large quantities of unwanted output can be both tiresome and distracting. We have, as far as possible, kept the default output terse and to the point. But, at the same time, directives are available that can produce, at the user's request, any other output that may be needed.

A final point on output - a 'transcript' file is now available. When switched on, a record of all input to and output from the program will be sent to this file, together with the relevant channel numbers. It is most useful for providing 'hard-copy' when working on a VDU.

3.6. *The HELP Facility*

GLIM-3 had a useful HELP facility, that provided more detailed information following a fault. It is hoped that an extended HELP facility will be available to give information on the choices open to a user both within a directive and between different directives. A 'feedback' facility is also planned whereby, when switched on, comments would be printed at the user's terminal, informing him/her of the actions taken by the program, with explanations as necessary.

4. THE PRISM LANGUAGE

In this section we describe several of the new language features of Prism -- optionwords and option-lists, prhasewords and phrase-lists, and clauses. First we begin with some definitions.

4.1. *Tokens and Lists*

Although it was never explicitly stated as such, there were, aside from the operators, three types of item used in GLIM-3 directives: (1) identifiers (2) values, such as numbers, integers, strings, and (3) pathwords which, as mentioned above, occurred as the names of choices in the ERROR etc. directives, although they never had such a name in GLIM-3 syntax. Identifiers, values and pathwords are the primitive objects of GLIM-3 (and Prism) language and we introduce the term 'token' as a generic name for all three. In Prism, tokens are the only items in many directives, form the operands in expressions, can be passed as actual arguments and can be stored in pointers and records (see section 5). In some circumstances, e.g. in argument-lists or in records, it is necessary to precede a pathword with the symbol $\sim$ to distinguish it from an identifier.

Lists occur frequently in Prism. The simplest is the token-list, which is a list of identifiers, values or pathwords. For example, the DATA statement contains an identifier-list. We note here that within any list in Prism the items may optionally be separated by commas, so that

$$\text{\$data} \quad 16 \quad \text{W, X, Y Z}$$

contains a valid list.

4.2. *Option-lists*

Now a feature of many languages (especially interactive languages) is the provision of 'signposts' to assist the user in declaring his/her choices to the System.

Signposts are usually implemented via option-lists: there is a set of choices, an optionword (or keyword) indicates which choice is being made and it is followed by the value chosen. Refinements include the ability to shorten the optionword and an implied ordering in the choices (i.e. the set is really a sequence) so that if an optionword is omitted the settings are assumed to be in serial order. All these features are present in the option-lists implemented in Prism. Optionwords (which are predefined by the System) always begin with a letter which may be followed by letters and/or digits and are **always** terminated by a '=' sign. Rules for shortening optionwords are similar to those for shortening directive-names in GLIM-3 (and Prism): at most four characters (excluding the '=' sign) are significant, but if fewer are used then the first match in the sequence, up to the number of characters used, will be taken.

Option-lists are used when, within a directive, a number of independent choices must be made, many of which will be defaulted. For example in the OUTPUT directive five choices have to be made and the optionwords and values (in correct sequence) are:

CHANNEL= integer
WIDTH= integer
HEIGHT= integer
ACCURACY= integer
WARN= integer

Commas are optional between option-settings in the list, and if choices are repeated then the last value set is used. Thus

$output   6, ACCURACY=3, W=72, , 2

sets the channel to 6, the width to 72 and the accuracy to 2.

4.3. *Phrase-lists*

Although option-lists are simple and convenient to use, they have the disadvantage that the choices are taken from an essentially linear sequence. Quite often this is too restricting, as when a nested sequence of choices is to be made, the options available at the next choice depending on the last setting chosen. For example, at the start of the DRAW directive of the GRAPH module it is necessary to select an action and one possibility is to select a style, the next choice being the type of style (point, line or area), and finally the style number must be given. In order to allow such choices to be made, in as natural a manner as possible, the concept of phrases (and the associated phrasewords and phrase-lists) was introduced.

A phrase consists of a phraseword (indicating the name of the phrase) followed by either a token, a formulae, a token-list, an option-list or another phrase; this latter ability gives the phrase its hierarchical structure. The phraseword itself has the same formation rules as the optionword except (and this is important) that it is never followed by an '=' sign. The phrasewords themselves have usually been chosen to give an English-like feel to the syntax; for example

$tabulate   the MEAN   of Y   by A.B   into TABLE

Here 'the', 'of', 'by' and 'into' are phrasewords, which are followed by, respectively, a pathword, an identifier, a classification formulae and another identifier. This sequence of 4 phrases forms a phrase-list, in which the phrases may of course be separated by commas. The phrases may be given in any order, and the last of any duplicate settings will always be used.

The hierarchical use of phrases is illustrated in the DRAW directive, as mentioned previously:

$draw   in   linestyle   3

where both 'in' and 'linestyle' are phrasewords introducing phrases. In most cases synonyms have been provided for the 'prepositional' phrasewords; these synonyms usually have a more immediate interpretation, as in the above example where 'style' and 'in' are synonyms, so that

$draw   style   linestyle   3

is equally valid.

4.4. *Clauses*

Finally, it is often useful to be able to subdivide a directive into 'clauses', each clause representing the description of one action, which will be performed before the next clause is read. The clauses themselves are separated by semi-colons. Directives usually have this structure when it is convenient that the settings used in one clause remain in force through subsequent clauses until reset. For example, the directive

$tabulate   the MEAN   of X   by A.B   into TX ;
                  of Y   into TY $

consists of two phrases: the first will tabulate the mean of X for each level of

AxB into the table TX; the second will tabulate the mean of Y for each level of AxB into the table TY.

5. DATA STRUCTURES

One of the more noticeable differences between GLIM-3 and Prism is the greater variety of data structures that are now supported. The casual user of the System may not notice the difference but for more sophisticated use the new structures should prove most helpful.

5.1. *The Different Types*

There are now six 'families' of data structures in Prism. These are:

scalar vector range text graphic referential

The first three are the numerical structures. The graphic structures are used only in the GRAPH module. The user interested in multidimensional arrays and tables will notice that there is no corresponding data structure for them: how they are handled is described in section 6.

Within each family of structures there are several different types, for example the vector family consists of the variates, the factors and the counts, these latter being new to Prism (see below). There are a total of 16 such types of structure and to each type there corresponds a directive in Prism for declaring an identifier to be the name of an instance of such a structure, to give it 'attribute' values and initialize it with data values. All such declaratory directives have a similar form: the directive name is followed by a list of declarations, each declaration consisting of an identifier, optionally followed by a list of 'attribute' values (preceded by '/'s), optionally followed by a data value (preceded by an '='). For example:

$variate   X/NR = (-2.8, 3.1),   Y = (*, 41.3E5, 92)   Z

declares X to be a variate with its first attribute value (see next section) as the identifier NR and initializes its length to 2 with the stated data values; it then declares Y to be a variate (with default attributes) of length 3 taking the given data values (including a missing-value); and finally it declares Z to be a variate with default attributes and zero length. Once an identifier is declared to be of a given type this cannot be changed, unless the identifier is first deleted and then reused. The attribute and data values may however be redeclared at any time, in which case settings not specifically altered remain unchanged. Another common feature is that a structure of any type may have its data values altered via the EDIT directive.

5.1. *Numerical structures*

The scalars, vectors and ranges are subdivided as follows:

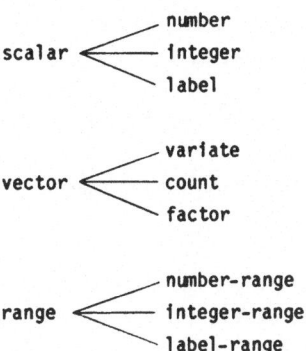

Each triplet within each family is used to store the 3 kinds of value that, in a broad sense, arise in statistical data. The first kind is the continuous measurement, such as distance, yield, time, etc., and is represented by the real numbers (known in Prism simply as 'numbers'). The second kind is the discrete or counting measurement and is represented by the non-negative integers. The third kind corresponds to membership of a finite set and in Prism is represented by a 'label', as described below.

A scalar structure stores a single value of the corresponding type, so a number structure stores a (real) number, an integer structure stores a (non-negative) integer and a label structure stores a label. A label denotes a member of a finite set, the set itself being defined by a label-range; a label is either an alphanumeric label in which case it begins with an underline symbol followed by a letter, then by letters, digits or underlines, or it is a numeric label in which case it begins with an underline symbol followed by digits; in certain circumstances the initial underline may be omitted. Examples of valid labels are:

_male   _67   _label   _set_1   _t22   _2

Scalars are used to store individual values, such as option-settings, totals, subfile labels, etc. Scalar structures are declared by the NUMBER, INTEGER and LABEL directives.

A vector is an n-tuple of numerical values, enclosed in parentheses and optionally separated by commas. A variate is an n-tuple of numbers, a count is an n-tuple of integers and a factor is an n-tuple of labels. Variates are stored in variate structures and similarly for counts and factors. Vectors are used to store sets of values, typically the observations from an experiment. Vector structures are declared through the VARIATE, COUNT and FACTOR directives. It is always permissible to follow the opening parenthesis of a vector by the initial letter of its type-name, to emphasis its type, as in:

$variate V=(v 4.2 5.6) $count C=(c 56 0 6 2) $factor F=(f _2 _1 _3)

and when, as in argument-lists, there is no other indication of the vector's type this initial letter is essential.

A range defines a set: number-ranges define (closed) subsets of the real numbers, integer-ranges define subsets of the integers, while label-ranges define sets of labels. A number-range consists of a list of number-intervals, each number-interval being of the form:

number-1 [# number-2]

where number-1 denotes the lower limit of the interval and number-2 denotes its size (taken by default to be zero). Thus the subset of the real numbers consisting of the point -23.2 and intervals [2.4, 4.7] and [-4.1, -2.0] could be stored as:

(-23.2, 2.4#2.3, -4.1#2.1)

Number-range structures are declared through the NRANGE directive, and the identifier NR used previously could have been declared by:

$nrange NR = (-23.2, 2.4#2.3, -4.1#2.1)

When used as the first attribute of a variate a number-range determines the set of valid values for that variate; it can be seen that the data values of the variate X above are within the range determined by NR, but the values 4.8 or 23, for example, would not be.

Similarly, integer-ranges are lists of integer-intervals and the corresponding structures are declared through the IRANGE directive. Label-ranges are lists of labels, whose structures are declared through the LRANGE directive; they may be used to define the set of permissible labels for a factor; for example

```
$lrange  GENDER = (_male _female)
$factor  SEX/GENDER = (_male, _male, _female, _female, _male)
```

Ranges are also used to name the dimensions of a table, and the levels within a dimension. They also have a role to play in converting between the different types of vector value and between the scalar types, but this will not be discussed further here.

5.3. *Text Structures*

GLIM-3 had only one text structure - the macro. It served two purposes, firstly to store sequences of directives and secondly to hold captions etc. This dual role often caused problems, as when a macro used in a WHILE stastement did not end with a $. In Prism, captions and other such text are stored in 'strings', which may not contain directive symbols. Macros now hold only directives (or more generally statements, see section 8), and begin and end with a $. Strings may be inserted within a directive, as in GLIM-3, while macros may appear in the USE, WHILE, SWITCH etc directives; see also section 8.

Strings are declared as in

```
                    $string  S = 'this is a string'
```
and macros as in

```
                    $macro   M = $print  S  $endmac
```

The attribute value for strings and macros is the static argument-list described in section 8.

A problem that occurs with the repeated use of macros in GLIM-3 (e.g. in the WHILE statement) is that the contents of the macro are reinterpreted on each execution of the macro. When the exercution time per directive is small this interpretation overhead can become a large proportion of total running time for the macro. Hence it would be desirable if macros could be translated to a more readily interpretable version, in much the same way that high-level languages are translated (compiled) closer to machine-code prior to execution. It is hoped that this facility will be available in Release 1 of Prism. These 'translated macros' are known as modules, the third type of text structure, and the translation would be achieved by the declaration:

```
                    $module   MOD = MAC
```

where MAC is a macro identifier. Once translated the module can be used wherever a macro can be used, but will execute more efficiently. Only in the EDIT directive will syntatic differences be noticed.

5.4. *Graphic Structures*

The graphic structures ('figures' and 'shapes') are of use only in the GRAPH module. A figure is a coordinate-free description of a two-dimensional graphical object (e.g. a line, a circle, a space); figures can be stored in figure-structures which are themselves declared through the FIGURE directive. Shapes are n-tuples of figures and are stored in shape-structures which are declared through the SHAPE directive. Shapes can be used to store and manipulate an abstract description of a 2-dimensional object. When matched with the data supplied by a record (see below) the object can then be drawn. Graphic structures are described in more detail by Slater (1982).

5.5. *Referential Structures*

Unlike the graphic structures, the two types of referential structures existed in an embrionic form in GLIM-3. %YV for example, unlike %FV, did not denote a variate in its own right, but 'pointed to' whatever was the current y-variate. In Prism terms %YV is a 'pointer' structure, the first type of referential structure: it points (or refers) to a token held elsewhere in the System, in the sense that should a pointer be used in Prism then the token that the pointer refers to will be used instead. Thus

$variate X $pointer P = X $look   P

will cause the values of X to be printed.

A record is an n-tuple of tokens, enclosed in curly brackets and optionally separated by commas. A record-structure may be declared by, for example,

$record REC = { X, (v 9.6,4.1), MA, ~NORMAL }

which declares REC to be a record holding 4 tokens: the identifier X, a variate value of length 2, the identifier MA and the pathword NORMAL.

Records have two main uses. Firstly, they hold the data values that will be used in the GRAPH module to map a shape onto the screen. Secondly they can be used to hold argument lists, as described in section 8.

6. ARRAYS AND TABLES

We take an array to be an n-dimensional (n>=0) structure whose dimensions are denoted by the integers 1...n, and whose levels within each dimension are denoted by the integers 1...l, where l is the size (or number of levels) of the dimension. We take a table to be an n-dimensional (n>=0) structure whose dimensions are named by the values of some label-range and whose levels within each dimension are named by the elements of some range. It can be seen that arrays and tables differ only in the naming of their dimensions and of the levels within each dimension.

The statistician whose data can be written as a 2-way table, as in

```
                        B
                  b1      b2
                 ---------------
         a1  |    10  |  14  |
     A                ----
         a2  |     5  |  19  |
                 ---------------
```

may enter the data into Prism by

    $factor A/(_a1 _a2) = (_a1 _a1 _a2 _a2), B/(_b1 _b2) = (_b1 _b2 _b1 _b2)
    $variate Y = (10 14   5 19)

As vectors the data will be in a form suitable for the GLIM, AOV or GRAPH modules. But the variate Y also has the form of a 2-way table classified by the factors A and B, and it would be convenient to be able to attach these 'attributes' to it to affect, for example, the way it is printed out. Some languages have a table or array type, but this can increase the number of structures the user needs to know about and can have the extra cost that the data may then have to be stored twice -- once in the vector structure and once in the table structure. Instead, in Prism, the user has the ability to attach table attributes to the original vector so that, when necessary, it can also be accessed, manipulated or printed as a table. This is set as the third attribute in the VARIATE, COUNT or FACTOR directives.

In fact the user can attach 2 different sorts of attributes to the vector. If a count is attached then the vector can be accessed as an array; the elements of the count give the number of levels for each dimension, the length of the count giving the number of dimensions. Alternatively, if a factor is attached then the vector can be accessed as a table; the label values of the factor (which must be distinct) both name the dimensions of the table and, when considered as identifiers, will be the names of ranges (or of vectors with ranges) to be used to name the levels of the dimension. For example, to declare Y as a 2-way array we would write:

$$\text{\$variate} \quad Y///(c \quad 2 \quad 2) = (\ 10 \ 14 \ \ 5 \ 19 \ )$$

while the table declaration would be written as:

$$\text{\$variate} \quad Y///(f \quad \_A \ \_B) = (\ 10 \ 14 \ \ 5 \ 19 \ )$$

It should be noted that counts and factors, as well as variates, can be given array and table attributes.

The array and table attributes are only used in the TABULATE, ACALCULATE, APRINT, TCALCULATE, TPRINT, PREDICT and OBTAIN directives. See Green (1982) for an extensive discussion of these directives and the general uses of arrays and tables.

7. DATA MANIPULATION

Rather than describe in detail the syntax and semantics of particular Kernel directives (this is already available in the Reference Guide) a brief summary will instead be given of the facilities available: in this section we describe the data manipulation directives, in the next the ones for program control.

The UNITS directive is still available, but its use is limited to setting a default length in the DATA statement. The individual declaration directives (e.g. the VARIATE, FACTOR, etc. directives) are available for setting values, or the DATA and READ directives may be used to set vector values in parallel; in this case it is now possible to ask for the length to be taken from the number of values found.

Throughout the System the length of a vector (or of a range, text, shape or record structure) is not considered fixed but may be changed at any time. In particular, a length of zero for these structures is taken as equivalent to missing value so that, for example, the following are equivalent:

$$\text{\$count} \quad C = * \qquad \text{\$count} \quad C = ( )$$

(But note that this is different to declaring C to have one, missing, value:

$$\text{\$count} \quad C = (*)$$

The EDIT directive is provided as a general facility for altering the data values of structures. The user may add, delete or modify the elements or lines of the structure, or the elements within lines where this is applicable. For simple calculations on vectors and scalars the CALCULATE directive has been retained but, to increase the efficiency of execution, only single operations are permitted. It is intended that most, general-purpose calculations will be performed through the ACALCULATE directive. The SORT directive is still available.

The LOOK command is used to display all the values of a structure; inspection of a subset of values must now be done through the SLOOK directive. The PRINT directive is more flexible than in GLIM-3, allowing considerable control over the formatting of output.

The PLOT directive has been retained, but can now be used in causal form, the scales for each plot being identical. It is also joined by the LPLOT directive, in which different subsets of the data will be plotted with different symbols. (Both directives are still intended for terminal or line-printer output.)

The ENVIRONMENT directive has been extended and a SET directive allows some of these System values (such as the batch/interactive mode) to be altered. A proposed SQUEEZE directive would perform additional garbage collection (beyond that usually performed by the System) to collect other available space.

8. PROGRAM CONTROL

Program control is concerned with the dynamic change of input sources; in Prism such sources may be input channels, macros or strings. Additionally when a change to a new input source is made it must be possible to associate arbitrary actual arguments with the predefined dummy arguments for that source.

Such a change is termed a 'redirection', of which there are 3 kinds: an input referral, a macro invocation or a string insertion. An input referral takes the form:

$$\text{channel } [\text{ width } [\text{ record }]]$$

a macro invocation takes the form:

$$\text{macro } [\sim \text{ record }]$$

a string insertion takes the form:

$$string [\sim record]$$

In each case the record holds the list of actual arguments for the redirection; 'channel' and 'width' refer to the input channel to be used, while 'macro' and 'string' refer to the macro or string to be used as the new source.

Input referrals occur in the INPUT, REINPUT, DINPUT and SUSPEND directives. For example,

$input   CHANNEL=15,   WIDTH=80,   RECORD= { X,   (f _a _b) }

switches control to channel 15 (with width 80) and supplies X as the first actual argument and the factor (_a _b) as the second.

Macro invocations occur in the USE, WHILE, SWITCH and DO directives; the DO directive, new to Prism, enables a macro to be invoked a fixed number of times. For example:

$do   5   M $\sim$ R

will invoke the macro M 5 times, each time with an argument-list as contained in the record R. Macros can also occur in 'direct macro invocations' which take the form: &macro-invocation. This has the same effect as a macro invocation via the USE directive but is more efficient and easier to use. In this case the   symbol preceding the record may be omitted. There are similarities between directives and direct macro invocations so that the word 'statement' is used to denote either of them. Assuming a macro MAC and other identifiers X, Y we could have for example:

$CALC   X = 3   &MAC { X, Y } $

String insertions may only occur as 'direct string insertions' which have the form:

string-insertion

For example:

#string S = '% log (%1)' $calc Y = #S $\sim${ X }

assigns Y the value of the log of X.

Macro invocations may be prematurely ended if, as in GLIM-3, an appropriate EXIT or SKIP directive is encountered. Input referrals are ended when a RETURN or FINISH directive is encountered. Input referrals may, as in GLIM-3, be directed to a particular subfile, but now the subfile is located via a label rather than an identifier.

The directives REWIND, DUMP, RESTORE and PAUSE are available as in GLIM-3, but DUMP only dumps as much of the workspace as is necessary, resulting in smaller binary files. The SAVE directive is new to Prism:

$$\$save \quad X$$

will write to the current output channel a Prism declaration such as would be needed to redeclare X to its present state.

Finally, argument-lists, as contained in records, have their own keyword system similar to that for optionwords, so that, for example:

$$\{ \quad \%3= Z, \quad \%1= X, \quad Y \quad \}$$

is equivalent to:

$$\{ \quad X, Y, Z \quad \}$$

The arguments-lists described so far are dynamic -- they apply only to the particular redirection for which they are used. Static, or default, argument-lists may be supplied as the first attribute of a MACRO or MODULE declaration and will be used unless overridden by a dynamic setting. Local structures are available by using the symbol '#' in place of an actual argument. Thus:

$$\$use \quad m \sim \{ X, \# \}$$

supplies X as the first actual argument and will cause the System to supply a local structure for the second argument and delete it at the end of the invocation. This allows for full recursion within redirections.

REFERENCES

ALVEY *et al*, (1977) *The Genstat Manual*. Rothamsted Experimental Station.

BAKER, R.J. and NELDER, J.A. (1978) *The GLIM Manual: Release 3*. Numerical Algorithms Group: Oxford U.K.

CLARKE, M.R.B. (1982) GLIM-4: the new facilities. See this volume.

GREEN. M. (1982) Array manipulation in Prism. See this volume.

HENSTRIDGE, J.D., PAYNE, R.W. and BAKER, R.J. (1982) Buffered output in Fortran. *Computer Journal*. To appear.

NELDER, J.A. and WEDDERBURN, R.W.M. (1972) Generalised linear models. *J. R. statist. Soc.*, A, *135*, 370-384.

PAYNE, R.W. (1982) AOV: the Prism module for analysing designed experiments. See this volume.

SLATER, M. (1982) The GRAPH module. See this volume.

WILKINSON, G.N. and ROGERS, C.A. (1973) Symbolic description of factorial models for analysis of variance. *Appl. Statist.*, *22*, 392-399.

GLIM4 - THE NEW FACILITIES

M.R.B.Clarke

Department of Computer Science and Statistics
Queen Mary College, University of London
Mile End Road, LONDON E1 4NS   U.K.

SUMMARY

GLIM is now one module of the new PRISM system.
The opportunity has been taken to completely rewrite
the program, improve the syntax and extend the range
of options. These changes are described and examples
of their use given. There are some incompatibilities
with GLIM3.

Keywords: GLIM; STRUCTURE FORMULA; LINEAR MODEL; MISSING VALUE.

1. INTRODUCTION

Unlike the GRAPH and AOV modules of the new system the GLIM
module is a development of what went before. Nevertheless the
opportunity has been taken to completely rewrite the program,
making the syntax consistent with the rest of the system,
improving existing facilities and in some cases extending them
quite substantially. There are several incompatibilities with
GLIM3 but it is felt that they are all justified by the more
consistent syntax and wider range of statistical methods that are
provided. This paper gives a summary of the main differences
between GLIM3 and GLIM4 together with some straightforward
examples of their application.

There are now several directives that can be followed by
structure formulae so the main changes in this area are described
first.

2. STRUCTURE FORMULAE

At the syntactic level the most obvious change for some users will be that the symbol 1 has been adopted (compulsorily) as a more elegant alternative to the previous rather clumsy $GM.

A more significant extension however is that factors can now have their levels coded by the user. Suppose for example that in the new syntax, Baker(1982) this volume, a factor is declared as

$FACTOR A/4=(2,4,1,3,4,2,3,1)

Then $FIT A sets up, as it always has done, a portion of the design matrix for A as

```
0  1  0  0
0  0  0  1
1  0  0  0
0  0  1  0
0  0  0  1
0  1  0  0
0  0  1  0
1  0  0  0
```

This can be considered as the default coding. The new program allows the user to specify a coding of his own with syntax exemplified by A<Q> where Q is a user supplied two-dimensional array having as many rows as there are levels of A and as many columns as he wishes there to be degrees of freedom for A. If Q for example is the array

```
1.1  2.2
3.3  4.4
5.5  6.6
7.7  8.8
```

then $FIT A<Q> generates internally the design matrix

```
3.3  4.4
7.7  8.8
1.1  2.2
5.5  6.6
7.7  8.8
3.3  4.4
5.5  6.6
1.1  2.2
```

Since a common requirement is for orthogonal polynomials over the levels of a factor there is a special facility to save the user having to set these up for himself. The syntax is A<n> where n is either an explicit integer or an integer scalar specifying orthogonal polynomials up to degree n. If A is as above then A<3> indicates the coding matrix

```
-1.5   1.0  -0.3
-0.5  -1.0   0.9
 0.5  -1.0  -0.9
 1.5   1.0   0.3
```

and the relevant portion of the design matrix after $FIT A<3> would be

```
    -0.5  -1.0   0.9
     1.5   1.0   0.3
    -1.5   1.0  -0.3
     0.5  -1.0  -0.9
     1.5   1.0   0.3
    -0.5  -1.0   0.9
     0.5  -1.0  -0.9
    -1.5   1.0   0.3
```

Supposing 8 to be the current number of units $FIT <Q> or $FIT <n>
are permitted alternatives for the frequently required

```
        $FACTOR I / 8⁻1
        $FIT I<Q> or I<n>
```

$FIT <X> therefore fits the user supplied design matrix X as
distinct from $FIT X which fits X considered as a vector variate.

Another orthogonal polynomial facility is X<n> where X is a
vector variate and n an integer or integer scalar. This gives
polynomials up to degree n orthogonal over the the values of the
variate X.

Products of variates are now allowed within a term, for example
$FIT X.Y , but X.X still reduces to X. The effect of the latter
can be achieved anyway by the numerically better X<2>.

A final detail is that exponentiation can now be used to
abbreviate large sets of terms. For example (A+B+C+D+E)**3 can be
used to specify all terms up to and including those of third
order.

3. FIT, TERMS and PIVOT

The purpose of the reintroduced **$TERMS** directive is to provide a
means for storing and updating the linear predictor structure
formula. It has no other effect. The default formula at the start
of a job is the single term 1 and thereafter formulae can either
be incremental,ie start with an operator, in which case the
stored formula is updated, or non-incremental in which case it is
overwritten. $TERMS $ resets the default and therefore has the
same effect as $TERMS 1 $.

$FIT followed by a non-null formula has the same effect on the
stored formula as $TERMS but goes on as before to fit the current
model However a change from GLIM3 is that $FIT $ now fits
whatever the stored formula is currently set to. This is
convenient if for example you have a macro that changes the
exponent in the power link and then fits whatever the current
model is.

Another incompatibility with GLIM3 is that the constant term, represented by the symbol 1, is now on the same footing as any other identifier, never being present by implication. This makes the syntax more consistent, simplifies to some extent the internal mechanism and also makes it easier to say what happens when formulae are updated.

Another familiar cornerstone of GLIM3 that has been refashioned is the standard output from the $FIT directive. This has been changed to show both the number of observations that contribute to the parameter estimates - necessary in the case of missing values - and also the change in deviance for incremental fits. An example of the new output is

```
$DATA 8 X $READ  1  2  3  4  5  6  7  8
$DATA 8 Y $READ  1  4  9 16 25 36 49 64
$YVARIATE Y
$FIT 1 $

   DEVIANCE = 3570.000
RESIDUAL DF =    7      FROM 8 OBSERVATIONS

$FIT +X $

   DEVIANCE = 168.000 (CHANGE = -3402.000)
RESIDUAL DF =    6      (CHANGE =    -1   ) FROM 8 OBSERVATIONS

$PROBABILITY DISTRIBUTION=POISSON $F$

SCALED DEVIANCE = 6.682 AT CYCLE 4
    RESIDUAL DF = 6       FROM 8 OBSERVATIONS

$FIT -X $

SCALED DEVIANCE = 150.675 (CHANGE = +143.993) AT CYCLE 4
    RESIDUAL DF =    7      (CHANGE =    +1   ) FROM 8 OBSERVATIONS
```

It is now possible in the $FIT directive to specify that some of the parameters implied by the stored formula are not to be fitted. Syntax is of the form

$FIT [sfa] [> sfb]

where sfa and sfb are structure formulae reducing to a set of terms such that sfb is a subset of sfa. The effect is that a triangle is set up for the terms in sfa but for those in sfb no parameters are estimated. $FIT > sfb $ takes the stored formula as sfa and there are similar rules for the other possibilities. The main use for this extension will be in conjunction with $PIVOT (below) for forward selection in stepwise fitting macros. In the non-iterative case it means that the unfitted vectors are on the same footing as the y-variate and the corresponding elements of the triangle can be used to select the next term to be fitted. This is usually done in the context of the sweep operator but, Clarke (1981), there is no theoretical reason why it should not also be done for orthogonal decomposition methods of the kind now to be used as the default in GLIM. More detail is in the section on numerical methods below.

The new directive **$PIVOT** [structure formula] will be provided to enable stepwise fitting to be specified. A simple example of its use is

```
$FIT U,V,W,X > V,W,X   $C fit U but not V,W,X
$DISPLAY TRIANGLE  $C examine residual sums of squares
$PIVOT X
$D T $    etc
```

The final syntax and most useful form of output for $PIVOT have not been decided at the time of writing so sample output is not given.

4. MARGIN and ELIMINATE

$MARGIN followed by a single term, such as $MARGIN A.B where A and B are factors, directs that until the next $MARGIN all vectors involved in a fit will be summed onto the specified margin. For example

```
$FACTOR A/2~3 : B/3~1
$DATA 12 X $READ 1.1  2.1  3.1  4.1  5.1  6.1
                7.1  8.1  9.1 10.1 11.1 12.1
$DATA 12 Y $READ 1.4  2.3  3.6  4.1  5.2  6.3
                7.7  8.3  9.4 10.2 11.5 12.3
$YVARIATE Y
$MARGIN A.B
$FIT 1+X $

  DEVIANCE = 0.154
RESIDUAL DF = 4     FROM 6 MARGINAL AVERAGES

$DISPLAY ESTIMATES,RESIDUALS $

          ESTIMATE      S.E.      PARAMETER
    1      0.5318      0.2262      1
    2      0.9586      0.0332      X
 SCALE PARAMETER       0.0385

MARGIN   AVERAGE   FITTED   RESIDUAL   MARGIN IDENTIFIER
   1      4.550    4.462     0.125     A(1).B(1)
   2      5.300    5.420    -0.170     A(1).B(2)
   3      6.500    6.379     0.171     A(1).B(3)
   4      7.150    7.338    -0.265     A(2).B(1)
   5      8.350    8.296     0.076     A(2).B(2)
   6      9.300    9.255     0.064     A(2).B(3)
```

The default is averages but totals - useful in the Poisson case can be specified (in the $ERROR directive). Note on the next page how the wording on the output is adjusted in this case.

```
$ERROR MARGIN=TOTAL $F$D R$

   DEVIANCE = 0.154
RESIDUAL DF = 4      FROM 6 MARGINAL TOTALS

   MARGIN      TOTAL      FITTED      RESIDUAL      MARGIN IDENTIFIER
     1         9.100       8.924        0.125        A(1).B(1)
     2        10.600      10.841       -0.170        A(1).B(2)
     3        13.000      12.758        0.171        A(1).B(3)
     4        14.300      14.675       -0.265        A(2).B(1)
     5        16.700      16.592        0.076        A(2).B(2)
     6        18.600      18.510        0.064        A(2).B(3)
```

Mathematically the operation of $MARGIN can be expressed as
follows.   Let [X|y] be the data and design matrix implied by the
current structure formula and y-variate before the $MARGIN and
let M be the design matrix corresponding to the margin formula.
Then until the next $MARGIN the design and data matrix used in
all fitting will be M'[X|y]. In the totals case above this is

```
               2     8.2      9.1
               2    10.2     10.6
               2    12.2     13.0
               2    14.2     14.3
               2    16.2     16.7
               2    18.2     18.6
```

In general any non-incremental formula can be supplied, but it
cannot contain coded factors and when expanded must only contain
terms marginal to a single high order term.   For example
$MARGIN A*B has the same effect as above but $MARGIN A+B is
meaningless. There is no algorithmic reason why the margin design
matrix has to be orthogonal but if it is not then of course the
marginal observations are not in general independent.

Inclusion of a variate such as $MARGIN A.B.X is permitted to give
weighted sums. The marginal values are neither explicitly formed
or stored and empty cells are ignored so that in general the
overhead of $MARGIN is one of time rather than storage.

$ELIMINATE [single term] has identical syntax but here the
effect is to eliminate the specified term from all terms
involved in subsequent fits. So that

        $ELIMINATE A.B $FIT X $

gives the same deviance and estimates for X as $FIT A.B+X $ but
the working triangle does not contain the possibly large block
corresponding to the A.B term. This is particularly useful for a
within group analysis or contingency tables with fixed margins.

```
$MARGIN $FIT A.B+X $
```

```
DEVIANCE = 0.094
RESIDUAL DF = 5      FROM 12 OBSERVATIONS
```

```
$DISPLAY ESTIMATES $
```

	ESTIMATE	S.E.	PARAMETER
1	1.0139	0.0132	X
2	0.3931	0.1111	A(1).B(1)
3	0.1292	0.1181	A(1).B(2)
4	0.3153	0.1261	A(1).B(3)
5	-0.0486	0.1349	A(2).B(1)
6	0.1375	0.1444	A(2).B(2)
7	0.0736	0.1545	A(2).B(3)
SCALE PARAMETER		0.0188	

```
$ELIMINATE A.B $FIT X $
```

```
DEVIANCE = 0.094
RESIDUAL DF = 5      FROM 12 OBSERVATIONS
```

```
$DISPLAY ESTIMATES $
```

	ESTIMATE	S.E.	PARAMETER
1	1.0139	0.0132	X
SCALE PARAMETER		0.0188	

In this case it is essential for the eliminated term to correspond to an orthogonal set, and the form allowed guarantees this. If the eliminated design matrix is E then in the notation above the data and design vectors used in the fit are the residuals $(I - E(E'E)^{-1}E')[X|y]$. If a margin is specified then E is replaced by $M'E$.

5. ERRORS and LINKS

The new keyword syntax is used here

$PROBABILITY DISTRIBUTION= INVERSEGAUSSIAN

showing the lengths to which it will now be possible to go, and also one of the three new distributions. The other two are a close approximation to the hypergeometric and a family for which the variance is of the form $\alpha + \beta u + \gamma u$ the negative binomial being a special case.

New links include a more general power link of the Box and Cox form $\eta=((\mu+\delta)^\lambda -1)/\alpha$ with α and δ being user supplied, and a composite link of the form $\eta=f[Cg(\mu)]$ where f and g are functions and C is in general a matrix.

With composite links it is possible for each y-variate element to correspond to several linear predictors. The syntax is not fully settled at the time of writing but certain forms of C corresponding to standard situations, a differencing operator for example,will be made available without the user having to calculate C explicitly. The composite link idea which it is hoped to further generalise in subsequent releases makes possible the analysis of contingency tables with combined cells and grouped normal data [Thompson and Baker(1981)].

6. MISSING VALUES

Missing values are now allowed in all the vectors that take part in a fit and are handled as follows.

A distinction is made between vectors involved in the linear predictor together with those associated with it namely those in the margin and eliminated formulae and the offset, call these lp-vectors, and those associated with the y-variate such as the binomial denominator, the initial value vector for the iteration and the prior weight; call these obs-vectors. The possibility of using composite link functions implies that several lp-vector elements may be associated with a single obs-vector element.

A missing value in any of these vectors results in all affected observations being omitted from the fit . A simple example is shown below

```
$DATA 8 X $READ  1  *  3  4  5  6  7  8
$DATA 8 Y $READ  1  4  9  *  25 36 49 64
$DATA 8 W $READ  1  1  1  1  1  0  1  0
$YVARIATE Y
$WEIGHT W
$FIT 1+X $

    DEVIANCE = 64.000
RESIDUAL DF =  2      FROM 4 OBSERVATIONS

$DISPLAY RESIDUALS $

    UNIT   OBSERVED   FITTED    RESIDUAL
     1      1.000     -3.000      4.000
    (2)     4.000       *           *
     3      9.000     13.000     -4.000
    (4)       *       21.000        *
     5     25.000     29.000     -4.000
    (6)    36.000     37.000     -1.000
     7     49.000     45.000      4.000
    (8)    64.000     53.000     11.000
```

Note that fitted values have been estimated both for the missing observed value and for the zero weighted observations. This can always be done for missing obs-vector values but of course is not possible in the case of missing lp-vector values. Note also the brackets round observations that did not contribute to the parameter estimates.

One consequence of the rules is that the data set can change from fit to fit depending on the pattern of missing values. It is to draw attention to this possibility that we now print out the number of observations that go into every fit. At the time of writing we are also considering the provision of a mechanism enabling the user to specify a superset of variates the intersection of whose complete data sets defines the set of units to be used in subsequent fits.

7. NUMERICAL METHOD

It would be possible to write a substantial paper on this topic alone and the view that the details should remain hidden from the user is an entirely reasonable one. Nevertheless the basic ideas can be quite easily explained. Consider a single iteration and ignore for simplicity the iterative weight. If the design matrix is X ,the observed response y and the parameters p familiar theory leads to the normal equations

$$X'X \; p = X'y$$

which can then be solved for the parameter estimates. In GLIM3 this is done by the Gauss-Jordan method which is often used in statistical programs because its intermediate results have a neat statistical interpretation and flexible stepwise strategies are easily implemented. Clarke (1982) gives the algorithm and details of how aliasing is implemented.

The problem with this method is that $X'X$ and subsequently $(X'X)^{-1}$ are explicitly formed and hence if X is at all ill-conditioned $(X'X)^{-1}$ can be very poorly determined.

A much better method is to perform some cancellation analytically in the normal equations by first expressing the design matrix in the form X=QU where $Q'Q=D$, a diagonal matrix, and U is upper triangular. The normal equations then reduce to

$$X'X \; p = U'Q'Q \; U \; p = U'D \; U \; p = U'Q'y$$

so that for the unaliased parameters

$$U \; p = D^{-1}Q'y$$

There are many ways of doing this. The popular Householder and Gram-Schmidt methods involve explicitly creating the design matrix which space considerations rule out in GLIM. Using Givens rotations however it is possible to restrict storage to exactly the same size triangle as required for the present Gauss-Jordan method while getting the benefit of improved numerical stability.

A single precision Givens algorithm is very nearly as accurate as double-precision Gauss-Jordan and it is becoming clear that it retains much of the latter's flexibility and convenient statistical interpretation; see Clarke (1980, 1981) for more detail.

Both methods have been made available in the new version and in addition the user will be given a choice of single or double precision at the terminal, be able to specify a further accuracy improving step known as iterative refinement, and if he has space to choose that the expanded design matrix is stored. This will save time particularly for slowly converging $OWN fits on complicated margins. In the new keyword syntax the full directive is

```
$METHOD ALGORITHM= GIVENS | JORDAN,
        PRECISION= SINGLE | DOUBLE,
             SAVE= SPACE  | TIME,
           REFINE=   YES  | NO
```

Of course the user with no need for these facilities need never
use the $METHOD directive. The default is single precision
Givens. To show however that it can make a difference consider
the following well-known cautionary example of an ill-conditioned
least-squares problem.

$DATA 8 X0 $READ 1 1 1 1 1 1 1 1
$DATA 8 X1 $READ 1 2 3 4 5 6 7 8
$ACAL X2=X1*X1 : X3=X1*X2 : X4=X1*X3 : X5=X1*X4 : X6=X1*X5
$ACAL Y=X0+X1+X2+X3+X4+X5+X6
$YVARIATE Y
$METHOD GIVENS
$FIT X0+X1+X2+X3+X4+X5+X6 $

   DEVIANCE = 0.000
RESIDUAL DF = 1   FROM 8 OBSERVATIONS

$DISPLAY ESTIMATES $

	ESTIMATE	S.E.	PARAMETER
1	1.0000	0.0000	X0
2	1.0000	0.0000	X1
3	1.0000	0.0000	X2
4	1.0000	0.0000	X3
5	1.0000	0.0000	X4
6	1.0000	0.0000	X5
7	1.0000	0.0000	X6
SCALE PARAMETER		0.0000	

$METHOD JORDAN F

   DEVIANCE = 0.000
RESIDUAL DF = 1   FROM 8 OBSERVATIONS

$DISPLAY ESTIMATES $

	ESTIMATE	S.E.	PARAMETER
1	1.0009	0.0009	X0
2	0.9981	0.0019	X1
3	1.0015	0.0015	X2
4	0.9995	0.0005	X3
5	1.0001	0.0001	X4
6	1.0000	0.0000	X5
7	1.0000	0.0000	X6
SCALE PARAMETER		0.0000	

In practice of course even on purely statistical grounds this
is not a sensible thing to do. It is precisely to handle this
situation that orthogonal polynomials are used. In this case
$FIT 1+X1<6> $ would be the appropriate statement.

8. OTHER DIRECTIVES

Full details will appear in the manuals and there is not the space for more than a brief summary here.

$DISPLAY and $EXTRACT have been integrated more closely. Any numerical structure that can be printed in $DISPLAY can be $EXTRACTed. The range of options in $DISPLAY has been greatly widened to reflect the considerably greater complexities of model and algorithm.

The user has been given full control over the tolerances that determine aliasing and convergence of iterative fits and any compatible vector can be specified to give the initial values for the iteration, the y-variate still being the default. $RECYCLE is now redundant.

An important new facility is $PREDICT which will enable tables of fitted margins to be produced. It has many similarities to $TABULATE, see Baker (1982) this volume, but operates on fitted values rather than raw data.

REFERENCES

Baker, R.J. (1982). PRISM - an overview. [this volume].

Clarke, M.R.B. (1980). Choice of algorithm for a model-fitting system. COMPSTAT 80. Proceedings in Computational Statistics,Edinburgh. Vienna: Physica-Verlag.

Clarke, M.R.B. (1981). Algorithm AS163. A Givens algorithm for moving from one linear model to another without going back to the data. Appl. Statist. 30,2,198-203.

Clarke, M.R.B. (1982). Algorithm AS178. The Gauss-Jordan sweep operator with detection of collinearity. Appl. Statist. 31 (in press).

Thompson, R. and Baker, R.J. (1981). Composite link functions in generalised linear models. Appl. Statist. 30,2,125-131.

ARRAY MANIPULATION IN PRISM

BY M. GREEN

Polytechnic of North London

SUMMARY.

The TABULATE directive for forming tables of summary statistics and directives for performing operations on multi-dimensional arrays give Prism a powerful new facility for data manipulation.

Keywords: ARRAYS, TABLES, TABULATE, ACALCULATE,
              TCALCULATE, PRISM.

INTRODUCTION.

The applications of GLIM have diversified in recent years to the analysis of sophisticated non-standard models and data manipulation either as a precursor to an analysis or to provide summary results to aid interpretation. In such applications matrices and multi-dimensional tables are of particular advantage. For this reason Prism provides facilities for the creation and manipulation of multi-dimensional arrays.

1. ARRAY SHAPE.

A variate, count or factor can be defined to be an array by attaching to it array attributes which define its shape. The shape specifies the number of dimensions and the number of levels for each dimension. In its simplest form a shape can be represented by a count variable with length corresponding to the number of dimensions and values the number of levels per dimension. The array attributes can be assigned in a declaration;

$$\text{\$COUNT   SHPE} = (2,3,2)$$
$$\text{\$VAR   x///SHPE}$$

or the shape can be changed within array expressions. The array attributes are used only in directives APRINT, TPRINT, ACALCULATE, TCALCULATE and FIT when specifying the design matrix. At all other times the array is treated as a vector.

2. ACALCULATE.

ACALCULATE is the general array manipulation directive and is followed by an array expression. This expression is of similar form to that in CALCULATE and wherever possible the result will be the same in both directives. Thus Y = %LOG(x) and A = B*C would have similar results. However, since no explicit looping is provided, many more operations have to be provided to make full array manipulation feasible. The syntax used is based on that of the APL language though not all APL facilities are available and some additional features are incorporated. In order to provide the large number of operators needed for a concise language,while restricting the character set to be as small as possible the operators are represented using double characters and combinations of operators. Wherever necessary operations are provided through functions. The benefits of a concise syntax far outweigh the initial difficulty the novice may experience.

2.1 *Basic Operations*

The monadic (unary) operators are + - and / (not). The "not" operator is used with the relational operators which return 0 for false and 1 for true. There are the usual mathematical functions,plus a few for dealing with missing values (see 2.7) The direct dyadic (element-by-element) operators are;
arithmetic: + - * / > < (minimum) < > (maximum) **
relational: < <= > > = == (equality) /= (inequality)
All have a functional equivalent, for instance %SUM(A;B) is equivalent to A+B. In general the operands must conform exactly in shape, vectors with no shape being taken to be 1-dimensional and scalars 0-dimensional. The exception is when one operand is a scalar which is taken to be an array conforming with the other operand and having all data values equal to the scalar value. Thus 2 * X has the obvious interpretation of multiplying each element of X by 2.

2.2 *Array Operations*

To allow more complex manipulation,operators for inner products, outer products and margining are defined by combining the direct dyadic operators with the prime character (1). If & and # are any two direct dyadic operators then the array operators have the following forms;

inner product	& ' #
outer product	' #
margin	& '

Inner and outer products have array operands while the second operand for margin is a count specifying the dimensions to be margined out. The direct dyadic operators specify the particular calculation to be performed. Thus in an inner product the element-by-element operation is defined by # and the subsequent "margining" by &. Matrix multiplication is thus a particular example of the inner product with the form + ' *. This gives great generality to these operations, for example the margins can be calculated as the products of the elements or an outer product formed using relational operators.

2.3 *Shape Manipulation*

The shape of an array can be accessed through use of the shape operator (υ). υA represents the shape of an array and is itself a 1-dimensionsl array and thus can be manipulatiod within an expression. An array's shape can be changed using the dyadic reshape operator ($\sim$). B $\sim$ S is an array with shape given by S and taking data values from B, repeating the values cyclically if necessary. The data values are ordered as last dimension varying fastest. Since data values and attributes can only be changed on assignment A = A $\sim$ $\wedge$ B is necessary to make A have the same shape as B. The shape of an array can be effectively removed by the unravel operator (,). Thus ,A is a 1-dimensional array with the same data values as A. This can be necessary when combining arrays using the concatenation operator (,). A,B results in an array containing the data values of A and B combined. Concatenation is on the last dimension, thus matrices are combined column-wise. In general all arrays in concatenation must have the same number of dimensions, with all but the last conforming with respect to the number of levels. However a d-dimensional and (d-1)-dimensional array can be concatenated, the last dimension of the (d-1)-dimensional array being taken to have one level. This allows scalars to be concatenated to "vectors" and "vectors" to "matrices". The result of a concatenation has the same dimensionality as the operands except for concatenation of scalars. Thus (1,2,3,4) has a 1-dimensional result.

The multi-dimensional generalization of transposition is provided through a permute function.

%PERMUTE (A;P) rearranges the data and attributes to put the dimensions in the order specified by P.

2.4 *Generating Data Values*

In order to be able to set the data values of an array or specify values as part of an expression some special operations are provided which produce a vector. Since an arithmetic progression is a common form for a set of data values a sequence operator (..) can be used in the following ways;

(i)   k,m..n
        numbers k to n in steps of (m-k)

(ii)   k..m,n
        numbers k to m in steps of n

(iii)   k..m
        numbers k to m in steps of 1 or -1

For example

(i)   2,5..14 = 2,5,8,11,14

(ii)   1.0..1.5, 0.1 = 1.0, 1.1, 1.2, 1.3, 1.4, 1.5

(iii)   9..5 = 9,8,7,6,5

Commonly data values occur in "blocks". Such data values can be generated using the multiple operator ('). Thus to have blocks of size 3 we use,

          3'(1,3,2,4) = 1,1,1,3,3,3,2,2,2,4,4,4

A more general form of this is to have a count vector as the first operand:

          (2,1,3)'(1.1,2.5,0.5) = 1.1,1.1,2.5,0.5,0.5,0.5

Any set of data can be made to repeat using the reshape operator,

$$2'(1..3)\sim 10 = 1,1,2,2,3,3,1,1,2,2$$

2.5 *Matrix Operations*

Four functions specific to 2-dimensional arrays are provided; %TRACE, %DETERMINANT, %SOLVE and %GSOLVE.

The last two take two arguments;

$$\%SOLVE(Y;B) \quad \text{returns} \quad Y^{-1}B$$
$$\%GSOLVE(Y;B) \quad \text{returns} \quad Y^{-}B$$

where Y^{-} is the generalized inverse of Y.

Inversion is not explicitly provided for reasons of numerical stability.

2.6 *Indexing*

The above operations are not powerful enough for many applications without the ability to operate on the elements separately. Individual elements can be treated as scalars by giving the index values, as in A(1;3;2;4). More powerfully, subsets of an array's elements can be specified through indexing. In this case vectors of index values are given, as in A(I;J;K;L). An indexed array is an array of the data values from A with indices in the subsets specified by the index vectors. For instance we may consider only those elements in certain rows and columns of a matrix. The index vectors may have shapes, the shape of the resulting array being the concatenation of the shapes of the index vectors.

Index values can be repeated in index vectors giving repetition of data values. An index vector may be omitted if all index values are to be used. Thus A(2;) refers to the second row of a matrix A.

When the subset of elements does not form a regular array functions %EXTRACT and %INSERT can be used. %EXTRACT (A;I;J;K;L) returns an array formed from elements of A. The nth data value of the result is the element of A with indices the nth values of the index vectors. The index vectors must have the same shape and the result has this shape. The complement to this function is %INSERT(A;I;J;K;L;B) which returns A with elements specified by I,J,K and L having been replaced by the data values of B.

2.7 *Missing Values*

As elsewhere in Prism a data value for an array can be a missing value and the missing value symbol (*) can be used within an expression as a value. In general operations involving missing values do not cause a fault, the result of any element-by-element operation involving a missing value will usually result in missing value (though missing value multiplied by zero has result zero). An invalid argument to a function or arithmetic operation will also result in a missing value, and a

warning message. Clearly missing value cannot be allowed in certain contexts and
will be faulted, for instance an array cannot be given a shape containing a missing
value. In order to allow the user to detect missing values and replace them,if
required,three special functions are supplied;

$$\%EQMV(X) = 1 \text{ (true) if } x = *$$
$$= 0 \text{ (false) otherwise}$$
$$\%MV(X;Y) = * \quad \text{if } y=1 \quad \text{(true)}$$
$$= x \quad \text{if } y=0 \quad \text{(false)}$$
$$\%VM(X;Y) = x \quad \text{if } x \neq *$$
$$= y \quad \text{if } x=*$$

2.8 *General Syntax*

The array expression can be as complex as required and allows assignment within
expressions. All operations have a priority and the operation of highest priority
will be performed first. In general evaluation is from left to right apart from
monadic operators, functions and multiple assignment.

3. USE OF LABELS WITH ARRAYS

A common form of multi-dimensional array in Statistics is the contingency table.
It is useful when displaying such an array to be able to label the dimensions and
levels within a dimension. This labelling can be specified through a label-range
or factor in the shape of an array. If R,S and T are factors then their label-
ranges will label the levels of the dimension of A on using,

$VAR A///(F   R,S,T) $

and the dimensions will have labels R,S and T.
I shall refer to such arrays as tables although there is no real difference between
an array and a table. An array without explicit labels is treated as having the
default labelling using the integers. A table output using TPRINT will display the
labelling.

4. TCALCULATE

TCALCULATE has the same operators as ACALCULATE but has some important differences.
Firstly a dimension or level must be referred to by label. If an array is used labels
_1,_2,_3, etc must be used. All shapes are factors,so that the second operand in
margining and permutation should be factors as should be index vectors.
In checks on conformity the labelling will be tested not only for correct length but
contents also. In direct dyadic operations the left and right operands will be made
conformable by permutation and expansion. Expansion is by repetition for a missing
dimension and by the insertion of missing values for missing levels.
This syntax relieves the user of the necessity of keeping account of the orders of

dimensions and levels after complex operations. The result of a TCALCULATE expression
will, of course, have the requisite labelling automatically.

5.1 *Tabulate*

The above directives give a powerful set of tools for manipulating arrays. However
raw data is often in a form that cannot be stored as a regular array. The TABULATE
directive is designed to deal with such data and provide a powerful method of data
summary.

The full syntax of TABULATE is;

```
$TABULATE    THE    TOTAL
                    MEAN
                    VARIANCE
                    MINIMUM
                    MAXIMUM
                    MEDIAN
                    PERCENTILE  value
input:       OF     vector
             WITH   vector
             BY     factor-list
output:      INTO   table          (statistics)
             AND    table          (weights)
```

  factor-list = vector [/subrange] [.vector [/subrange]]s

The action is to form summary statistics (defined by THE phrase) of observations
(OF phrase) weighted (WITH phrase) if required classified by various factors
(BY phrase). The observations within a classification will be used to calculate the
statistics. A missing value in the observation, weight or factor will result in the
observation being ignored. This makes it especially necessary to be able to have
access to the number of observations contributing to a particular statistic. The
values stored in the table specified in the AND phrase will, in general, be the
total weights. In the case of TOTAL and VARIANCE they will be a suitable weight for
that statistic.

The resulting tables will automatically be given the necessary labels and can be
treated as tables or arrays afterwards.

The factor-list can be very much more complicated than suggested above. If not all
levels of a factor are required a sub-range can be supplied specifying the levels to
be included;

```
            $TABULATE THE MEAN OF WAGE BY SEX/_MALE. CLASS
            INTO   TABL $
```

would form the (unweighted) mean wage for males only for each level of CLASS and
store the result in TABL.

5.2 *Grouping*

In many cases we have a variate or count which is required to produce a classification.
A grouping can be defined using a subrange of the form;

(value [# extension] [,value [# extension]]s)

A data value falls in a particular group if it lies between (value) and (value +
extension) for that group. A zero extension can be omitted.
Overlapping groups are rationalized to produce disjoint grouping.

AGE/ (16 , 20# 6, 25 # 5)

could be used instead of a factor to specify a classification by AGE grouped into

16 , (20-25), (25-30)

Any phrase not required may be omitted. In particular if no OF phrase is used
observations equal to one are assumed. This has particular use in forming contingency tables since

$TAB THE TOTAL BY SEX.CLASS INTO TABL $

will form a table of frequencies.
The TABULATE directive combined with TCALCULATE expressions together provide
powerful tools for data manipulation either as a precursor to an analysis or for
providing tables of summary statistics for interpretation of results.

THE GRAPH MODULE

M. Slater
Department of Computer Science and Statistics,
Queen Mary College, (University of London),
Mile End Road,
LONDON E1

SUMMARY
The main features of the GRAPH module of Prism are described. Details
of the main directives are given together with some examples of their
use. The relationship between GRAPH and GKS is discussed.

KEYWORDS Computer graphics; GLIM; GKS

1. INTRODUCTION

GRAPH is an interactive computer graphics program driven by a command
language similar in style to GLIM and AOV (Baker; Clarke; Payne,
1982). It is interactive in the dual sense that commands to GRAPH are
given at the terminal rather than via a batch program, and second,
pictures may be created interactively by the use of a graphical input
device (such as a light pen). GRAPH is modelled on and implemented via
the Graphical Kernel System (GKS), a functional description of a
graphics language which is the draft ISO standard on computer graphics
(Hopgood, 1982).

GKS (and hence GRAPH) supports certain distinctive features: first,
several graphical devices may be driven concurrently instead of one at
a time; second, device independent segment storage is provided, which
means that pictures may be saved, transformed, and redisplayed on any
device, irrespective of the storage capabilities of that device;
third, graphical output files are supported, which means that a
description of a picture may be saved in an external file and in a
subsequent session read back into the system redisplayed on the
devices in use, and perhaps edited.

GRAPH itself has capabilities not directly available as GKS functions
- in particular higher level graphical constructs (figures) are
predefined, and the user may construct concatenated sequences of such
figures (called shapes) to produce complex structures in a simple
manner. Standard shapes are predefined which are of special interest
in data analysis, such as scattergrams, histograms, contour plots, and
so on.

Since GRAPH is a module of Prism, all the utilities provided in the
kernel are, of course, also available. The array calculation
facilities will be of special importance to GRAPH users. Moreover,
within a Prism session GRAPH, GLIM and AOV directives may be used at
any time, making the overall system extremely powerful as a tool for
s atistical modelling and analysis. The implementation structure of
GRAPH is shown in Fig. 1.

2. BASIC DRAWING

All drawing in GRAPH is accomplished in the last instance by means of
four output primitives. These are **points**, **lines**, **areas** and **text**. 'In
the last instance' because even if the user invokes a higher level
function to draw, for example, a contour diagram, it is these four
output primitives which are invoked internally (in fact by calls to
GKS functions).

Every output primitive has a corresponding **style.** For example, lines
may be dashed or solid. Areas may be shown in outline (hollow) or

44

shaded in some manner. Points may be represented by dots or stars or
other symbols. Text may be drawn in different fonts and sizes.
Finally, colour is a feature of every output primitive.

The style of the primitive is represented by an index number which is
a pointer into a table of style chacteristics. For example, for lines,
the appropriate characteristics are the type of line (solid, dashed,
etc.), the width of the line, and the colour. Hence, line style index
number 1 may refer to red solid lines of standard width. Area index
number 2 may refer to areas filled in blue with a certain hatch
shading. The style index in force for each primitive may be changed by
the user at any moment in a GRAPH session. In addition the meaning of
each style index number may be defined and redefined by the user.
However, default settings will be in force which the user may leave
unchanged.

A seemingly complicating but actually very useful feature is that the
same style index number (say, linestyle 1) may be defined to have
different meanings on different graphical devices. Suppose the user is
sending graphical output to two devices, a colour display and a
plotter with one (black) pen. It might be useful to represent
'colour' on the plotter by different linestyle, and different colour
shadings by different hatch styles. So perhaps linestyle 1 may mean on
the plotter a dashed line, whereas on the colour display a solid blue
line. Linestyle 2 on the plotter might mean a dotted line
corresponding to red lines on the colour display.

To summarize, drawing is accomplished by means of four output
primitives; output primitives have styles which are pointers into
collections of characteristics on the graphical devices.

These concepts are realised by the directives $DRAW, $STYLE and
$DSTYLE. $DRAW accesses the output primitives. $STYLE sets the style
indices. $DSTYLE (Device STYLE) defines the style indices on the
devices. A discussion of $DSTYLE is deferred to section 3.

Example 2.1
$C Draw 4 lines criss-crossing the display area, each in
      four different styles

$STYLE lines 1 $DRAW from (0,0); to (1,1)
$STYLE lines 2 $DRAW lines (0,1), (1,0)
$STYLE lines 3 $DRAW poly line (0.5,0.5) (0,1)
$STYLE lines 4 $DRAW from (0,0,5); by (1,0)

Note the different methods of line drawing:

$DRAW from point; to point;   !absolute positions
     from point; by point;   !relative postions
     lines point-1,point-2,...,point-n;
     poly lines x y   !array of points
                                     ! given in x and y vectors

Note also that the example may be written more succinctly as

$DRAW using lines 1; lines (0,0) (1,1);
     using lines 2; lines (0,1),(1,0);
     using lines 3; lines (0.5,0),(0.5,1);
     using lines 4; lines (0,0.5),(1,0.5)

so that styles may be reset during the $DRAW, although on exit from this directive styles revert to their previous settings.

3. DEVICES

A device in GRAPH is a system which 'receives' graphical output, or which sends information back to the GRAPH program. Usually (but not essentially) the device corresponds to a physical graphical display machine with a display area such as a screen or paper. However, a device in GRAPH may be a form of storage which receives graphical data or returns information back to the program.

A device is identified to the system by two preset (but implementation dependent) integers. The first, called the CHANNEL, informs the GRAPH system of the channel number down which data is to be sent in order to reach the device. The TYPE is a number which identifies the type of device being referred to - eg, a certain tektronix model, a calcomp plotter, or a non graphical device such as the Segstore to be discussed later. The user must know the channel and type of each device to be used in the GRAPH session. (It is possible of course to inquire this information from the system). The user will choose an (integer) identifier for each device, and GRAPH will remember that this user supplied name for the device is a reference to a certain TYPE and CHANNEL.

A device may be in one of three states: **closed, open** or **active**. A device is by default closed, that is, no communication is possible, and no user identifier has been established. If a device is closed the user may **open** it, at which point the user name must be provided. When a device is opened inquiries may be made concerning its style settings, size of display area, etc.; also the user can redefine style settings, and receive graphical input. However, output primitives cannot be sent to the device until it has been **activated**. Unlike most graphics systems, there is in principle no limit to the number of devices which may be opened or activated at any given time. Note that output primitives are sent to all active devices. Putting this another way if no devices are active, $DRAW will not work.

$DEVICE is the directive for achieving these settings. Suppose two devices are to be used, a plotter and a colour device (this example will be followed throughout the rest of the paper where the colour device will be called a sigma).

$INTEGER sigma = 1 : plotter = 2

$DEVICE sigma **open** CHANNEL=6,TYPE=9; **activate**
  :   plotter **open** 4,4
$DRAW **from** (0,0); **to** (1,1) !line is drawn only on sigma

$DEVICE sigma **deactivate**
  :   plotter **activate**
$DRAW **from** (0,1); **to** (1,0) !line is drawn only on plotter

$DEVICE sigma **activate**
$DRAW **poly area** (0.2,0.4,0.3) (0.2,0.4,0.6)
                                !triangle is drawn on both

Subsequent output is directed to both devices, until they are deactivated.

Device styles can be set once a device has been opened, by

```
$DSTYLE sigma lines 1, TYPE=solid, WIDTH=1,
                          COLOUR=1;
               colour 1, RED=1.0, GREEN= 0.0, BLUE=0.0
       :     plotter lines 1, TYPE=dashed,WIDTH=2
```

Hence on the sigma, line style 1 means a solid line of standard width
(1*the predefined nominal width for this device type). Each device
has a RGB colour table, and on the sigma, colour index 1 is red. On
the plotter line output drawn while linestyle 1 is in force will be in
dashed lines of twice the nominal width.

Fig. 2 summarizes the discussion of the last two sections.

4. COORDINATE SYSTEMS

The choice of coordinate system representations follows logically from
the structure of GRAPH/GKS as a multi-device system. GRAPH users would
normally wish to define and use their own coordinate systems. In
addition each graphical device will have its own coordinate system as
determined by the manufacturer. Finally, there must be a standard
system for representing the display areas independently of any user or
device coordinate system. These requirements naturally lead to three
types of system: World Coordinates (WC) are the user coordinates;
Normalized Device Coordinates (NDC) is the unit square with corners
(0,0), (1,1) which is the standard device independent system. Finally,
there is a Device Coordinate (DC) system for each graphical device.

Suppose the user works in the coordinate system $-100<x<100$,
$-100<y<100$. Consider

```
$ACAL theta = 0..(2*%PI),0.05
   :    x    = 20.0*%COS(theta)
   :    y    = 20.0*%SIN(theta)
$DRAW poly line x y
```
(i.e., a circle of radius 20 at the origin)

Internally the user's coordinates are converted into NDC coordinates.
These NDC coordinates together with the instruction to draw are sent
to each active device, and each device driver converts the coordinates
into local DC.

User control is facilitated since it is possible to select a
rectangular region in NDC to which the WC system is to be mapped.
Bearing in mind that a particular user coordinate system is called a
window and the region of NDC space to which the window is mapped is
called the **viewport**, consider

```
$GMAP 1 window (-100,100,-100,100);
        viewport (0,0.5,0.5,1)
```

This defines a WC->NDC mapping so that the point $(-100,-100)->(0,0)$
and $(100,100)->(0.5,1)$. The circle would have appeared in the top left
hand quadrant of the display area had this mapping been in force. A
number of such window/viewport mappings may be defined by the user.
(By default all windows and viewports are the unit square). To select
a particular mapping,

```
$MAP i
```

where i is the integer representing the map. It should be noted that
the window/viewport mapping is not uniform, so that unless the window

and its corresponding viewport have the same aspect ratio circles will appear as ellipses.

The $GMAP directive defines Global MAPs, i.e., global to all devices. However, it is possible to arrange that on each device only a subwindow on NDC is displayed and mapped uniformly to the screen display area (which itself may be redefined by the user). With vectors x and y defined as above, and assuming that map 1 as defined is in force,

$DMAP sigma subwindow (0,0.5,0.5,1)
   :   plotter subw (0,0.5,0.75,1)
$DRAW poly line x,y

would produce the results shown in Fig. 3.

Note that NDC space is a 'logical' display space - the full picture on NDC may or may not be actually displayed on any device, depending on the subwindows in force. The default subwindow is the whole of NDC. The screen display area can be changed by

$DMAP sigma screen vector

where the vector is of length 4 giving a rectangular region of the display space in DC. This section is summarized in Fig. 4.

5. PICTURE SEGMENTS

Once an output primitive has been sent to a device it is lost to the system - it cannot be redrawn or transformed. Picture segments are a means of saving a sequence of output primitives (together with the styles in which they were drawn) as a collection identified by a common name (the segment name). Segments can then be manipulated as a whole - redrawn, transformed (i.e., scaled, rotated and shifted relative to a fixed point), made invisible or visible, and even inserted into another segment.

Some types of graphical device (refresh displays) have the capability for saving such picture segments locally. Moreover, since on such devices the picture is refreshed at least 30 times a second, if a segment is, for example, transformed, the effect will shown instantaneously thus allowing for animation.

The inexpensive and more popular graphical displays (e.g., rasther devices) do not have this capability. Hence GRAPH provides a device called the Segstore in which segments may be saved and displayed on any device. On such a non refresh display, if a segment is transformed, the effect will not be apparent until the picture is updated. When this happens all segments are redrawn with their most current characteristics. Similarly, if a segment is deleted or made invisible, no effect will show until the picture is updated, when the deleted or invisible segment will not be redrawn.

Once a segment exists (i.e., has been named and opened by the user) it will be in one of two states, open or closed. While a segment is open all output primitives are collected into it. Once the segment is closed, no further primitives can be added into it - it cannot be reopened. A closed segment can, however, be inserted (under a transformation) into the open segment. Only one segment can be open at a time.

The Segstore device is opened and activated as any other device. It should be noted that both the Segstore must be active and a segment open for output primitives to be collected into the Segstore. In addition a segment will exist on every device that was active at the time of its creation. It is, however, possible to copy a segment from the Segstore to another device.

Example 4.1
```
$INTEGER seg = 3
$DEVICE seg open CHANNEL=2, TYPE=0; activate
$SEGMENT open 1
  $DRAW poly line x,y;
        using lines 2; the border
$SEGMENT close
```

The circle, and a border around the current viewport are displayed on all active devices. They are saved in segment 1. Segment 1 exists on all active devices - in particular in the Segstore.

$SEMAP 1 AT=(0,0), SCALE=(0.5,0.5), SHIFT=(2,2)

will scale the segment by 0.5 for both X and Y axes relative to the origin, and shift the segment by the amount (2,2).

$PICTURE sigma update

will update the picture on the sigma, which entails clearing the screen and redrawing all the segments created while the sigma was active. When segment 1 is redrawn its new mapping will be taken into account.

6. GRAPICAL FILES

The Segstore only exists for the duration of the GRAPH session, it cannot be used as long term storage (between sessions). However, a graphical output file can be created. This is another device which keeps track of all GRAPH commands and output primitives and saves them in a file.

```
$INTEGER gfile=4
$DEVICE gfile open CHANNEL=20, TYPE=1
```

opens the output file (channel number is implementation dependent). Once the file is open it keeps a record of global settings (such as maps and styles), and may have style interpretations written into it. For example,

$DSTYLE gfile lines 1, TYPE=dashed,COLOUR=blue

will result in this information being written into the file. When the file is read back into the system, this interpretation for linestyle 1 will be set on all devices active at the time.

$DEVICE gfile activate

will activate the file, so that output primitives (and segments) will be stored on the file.

Once a file has been created and closed it can subsequently be used as a graphical input file which is treated as a special (input) device.

```
$INTEGER ginfile = 5
$DEVICE ginfile open CHANNEL=21, TYPE=2
```

opens the device (it cannot be activated because it is not an output device!) and

```
$GINPUT ginfile
```

would read the file contents and route the commands to active devices.

7. GRAPHICAL INPUT

Interactive graphics means the capability of interacting with a graphical device, that is not just sending data to it, but a two way traffic. For example, locating a point on the display area, returning the coordinates of that point to the program, locating a second point and then maybe connecting the two points with a line. In this way pictures can be constructed interactively, without the necessity of the programmer working out the coordinates of the points which make up the picture. In fact the coordinate system becomes irrelevar : in such work. (Most of the figures in this paper were constructed in this way).

There are many different types of input mechanism. The most well known are light pens, keypads, joysticks and mice. GKS is concerned not with the actual physical implement used, but the function to be performed. Hence GKS (and GRAPH) defines 'logical' input mechanisms to perform certain functions irrespective of their physical embodiment.

There are five classes of input mechanism defined by GKS: locators (to return coordinates); valuators (to return a real number); pick (to return a segment name); choice (to return an item on a menu); and string (to return a string). There are three modes of interaction defined called request, sample and event. A request interaction takes place in response to a user action such as depressing the hit button on a key pad. This is the only interaction considered here as the present release of GRAPH provides only request locator input.

A locator mechanism is first initialised by using the $DEVICE directive:

```
$DEVICE sigma locator 1, ECHO=on
```

This informs the system that a locator mechanism on the sigma device is to be switched on. There may be several locators connected to a graphics device, hence the particular one required is identified by the appropriate index number.

```
$REQUEST sigma locator 1,
             points list-of-points,
             STATUS=i, MAP=j
```

will return the points in the coordinate pairs given in the list of points. The STATUS will be returned as $i = 1$ if the hit was successful, otherwise $i = 0$. The mapping with highest priority in which the point was found is returned as the value of j. It should be realised that viewports may overlap, so that a point on the display area may be in several viewports. However, the user may set up a priority list, and the point is returned in the world coordinates of the map with the highest priority in which the point is located.

8. FIGURES AND SHAPES

GKS is a kernel system - it provides a set of minimal and orthogonal functions on top of which rather more flexible and user friendly programs can be built. GRAPH is broader than GKS in the sense that it provides the user with functions which are packaged sets of calls to GKS routines. For example, the user of GRAPH will not have to draw contour diagrams by direct use of the output primitives. Such higher level constructs are made directly available by the system.

The user is provided with a set of elementary figures (rectangle, ellipse, line, etc..) which can be used as building blocks for more complex structures. Following the ideas of Shaw (1969) and Barth et. al. (1981) every figure has a Head and a Tail. Figures can be concatenated together to form higher order structures called shapes, since the H of one figure is joined to the T of the next, for all figures in the shape. A shape therefore also has a H and T, its T is that of the first figure in the shape and its H is that of the last figure. Thus a figure is a standard arrangement of output primitives, and a shape is a concatenated sequence of figures (and shapes). Shapes are given a user identifier when created, and are referred to by that identifier until deleted. It should be noted that when a shape is defined nothing is drawn. A shape is merely a template that might be drawn at the appropriate time.

$DRAW the shape-identifier data-record

will result in the shape being drawn. The data record consists of data (in WC) appropriate for the figures and shapes defined in the shape referenced by the identifier. The data record will contain vectors (appropriate to the figures in the shape), and data records (appropriate to the shapes within the shape). If the data record provided is longer than that needed by the shape being drawn then the pointer into the shape moves back to the start and continues executing until the data record is exhausted. It can be seen that quite complex structures can be built using these facilities.

Axes, histograms, contours etc., are in the context of GRAPH standard shapes. Users can use the shape definition directive to define their own version of, say, a histogram (which is after all only a concatenated sequence of rectangles). Alternatively, the standard shapes may be used for convenience and efficiency.

9. TEXT

Text output might require a paper in itself to do it justice. GRAPH follows GKS text very closely. Text is an output primitive which therefore has an associated style. For example,

$DSTYLE sigma text 1, FONT=1, PRECISION=best,COLOUR=1

would define for the sigma textstyle 1 as given. In addition it is possible to specify in a device independent manner the geometrical properties of text output. Consider,

$GSTYLE 1 characters
          PATH=right, SPACING=0.0,
          HEIGHT=1.0,EXPANSION=1.0,
          UP=(0,1)
$STYLE characters 1

This requires that (on all devices) the text path will be at right

angles in a rightward direction to the up direction given by the UP
vector (which in this example is vertical). There is to be no
additional spacing between letters, the height of the characters is 1
unit (in WC), and the ratio of width to height of the characters is as
defined in the font (expansion=1). On all devices where precision is
best text must appear exactly as defined in the current character
style. However, if precision is defined as **ok**, then hardware
characters may be used and the character style ignored. Some examples
of text output are given in the appendix.

10. CONCLUSION

At the time of writing GRAPH is still in the development stage.
However, even at this relatively early stage its power is apparent.
Not even its creators (R.J. Baker of Rothamstead Experimental Station
and myself) have explored completely the potential of the being they
have conjured - especially when used as a tool of data analysis in
conjunction with GLIM. For example, at the simplest level consider a
regression analysis of Y on X. The user could set up 4 viewports
corresponding, say, to the four quadrants of the display area. In one
viewport draw the scatter diagram of Y on X. Use GLIM to obtain the
fitted values and residual. In a second viewport display the fitted
model together with the observed points. In the third viewport display
the scatter plot of residuals against fitted values. In the fourth
viewport display some text describing maybe the origins and definition
of the data, and conclusions reached. Save each of the four viewports
and contents in separate segments, and edit the the overall picture by
segment transformations until satisfaction is achieved. Write all the
segments to a hard copy plotter using the scale required. More complex
examples will no doubt come readily to mind.

REFERENCES

Baker, R.J. (1982) The Prism Kernel, Proceed. GLIM 1982,
    The North London Polytechnic, Springer.

Barth, W., Dirberger, J and Purgthafer, W (1981)
    The High-Level Programming Language PASCAL/GRAPH,
    Eurographics 81 (North-Holland)

Clarke, M.R.B (1982) The GLIM-4 Module, Proceed. GLIM 1982

Hopgood F.R.A., (1982) GKS - The First Graphics Standard,
    (Science and Engineering Research Council).
    Graphical Kernel System,
    Draft International Standard, ISO/TC97/SC5/WG2/N117

Payne, R. (1982) The AOV Module, Proceed. GLIM 1982

Shaw, A.C. (1969)
    A Formal Picture Description Scheme as a Basis for Picture
    Processing Systems, Information and Control 14

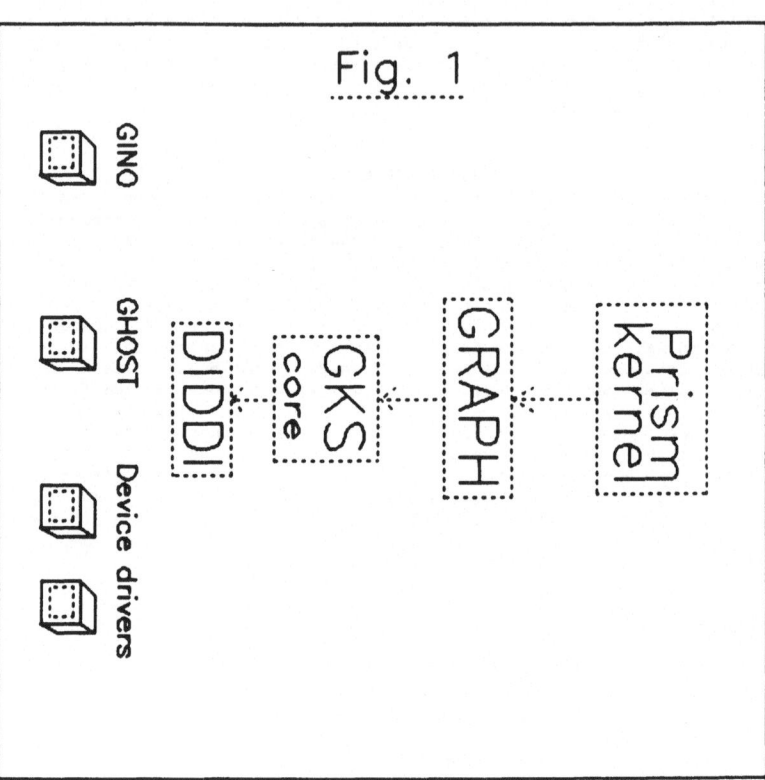

Fig. 1

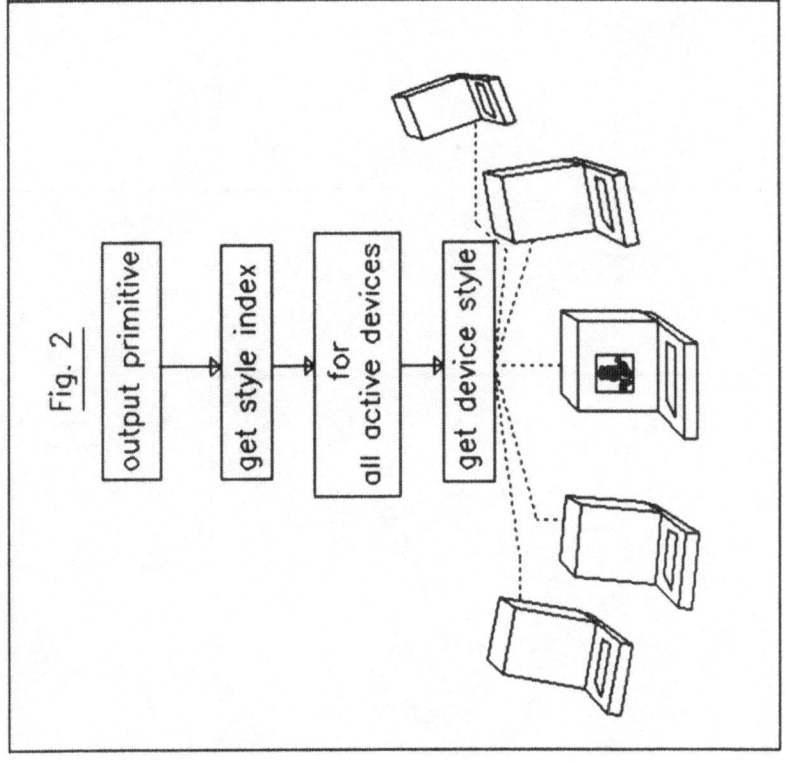

Fig. 2

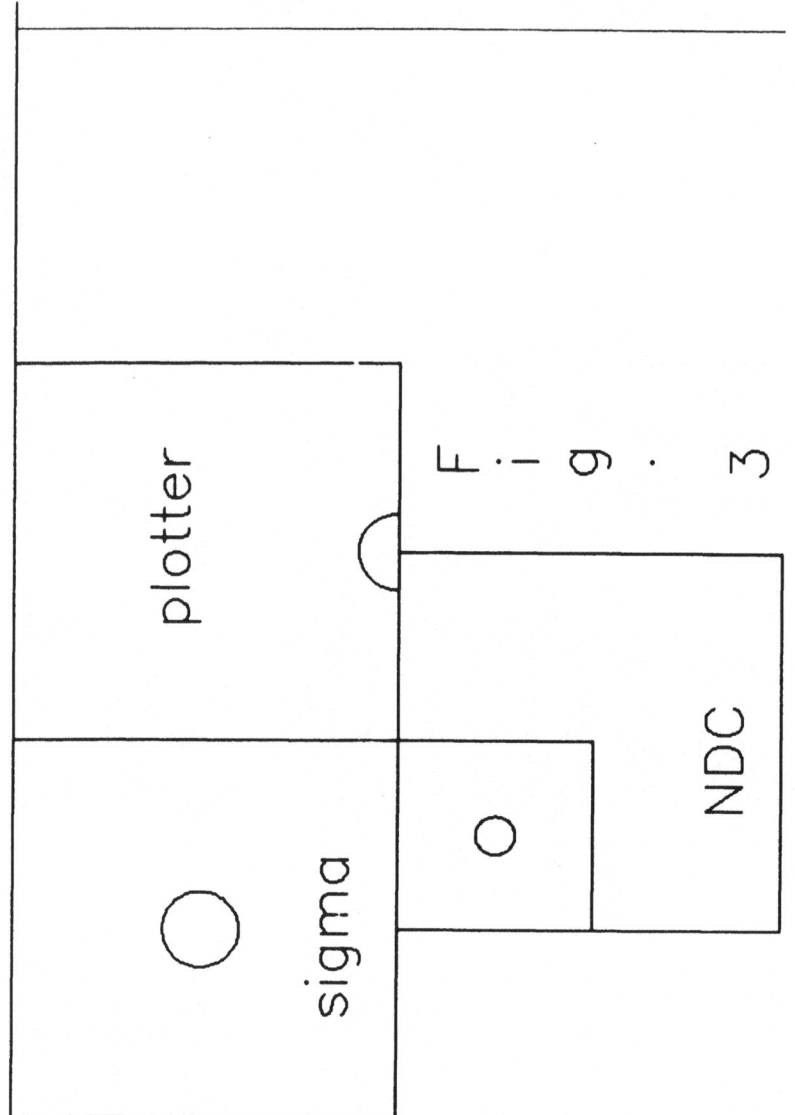

Fig. 3

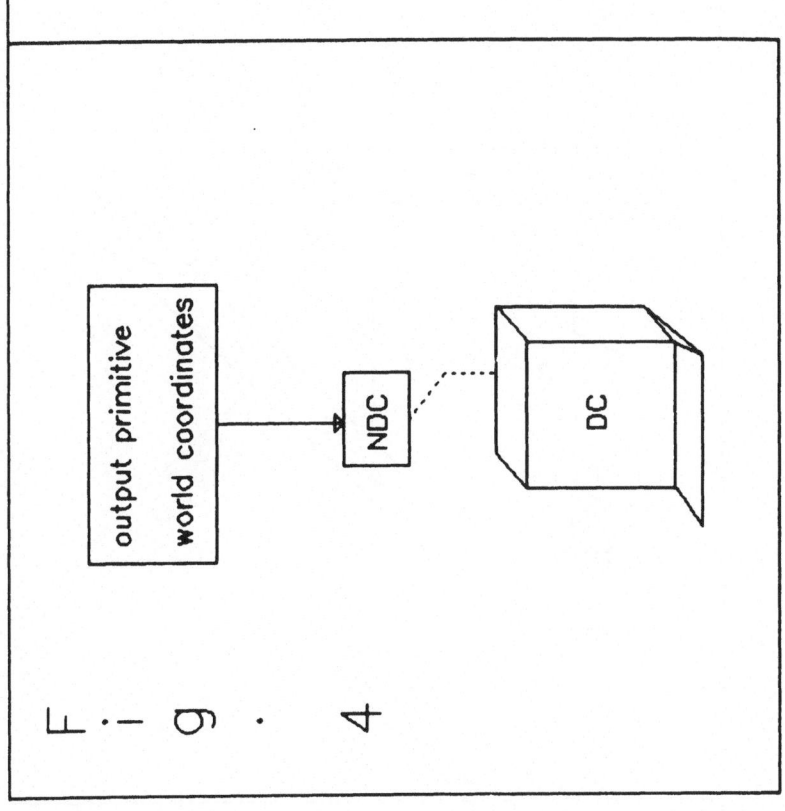

Fig. 4

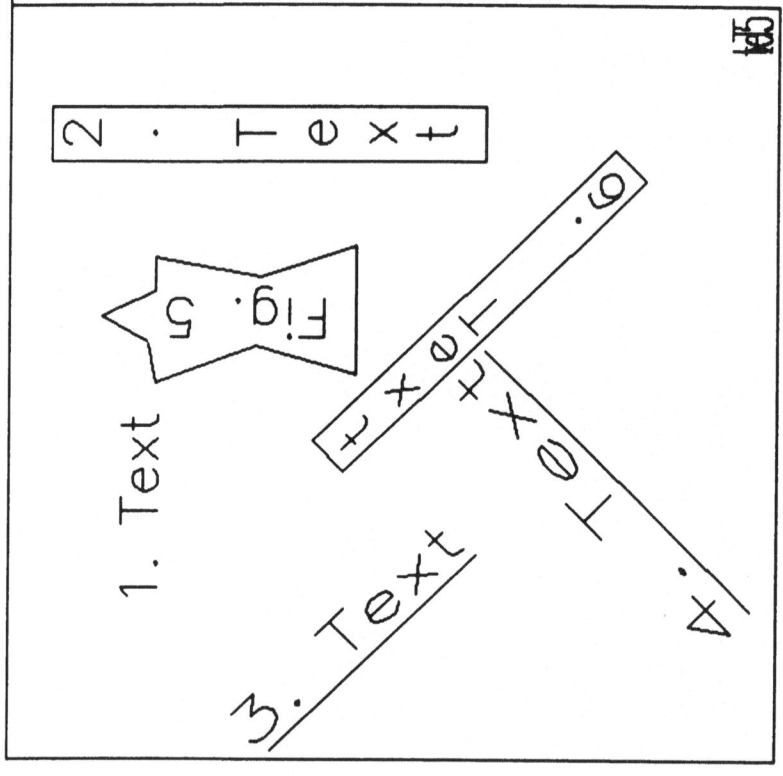

2 . Text

Fig. 5

1. Text

6. Text

4. Text

3. Text

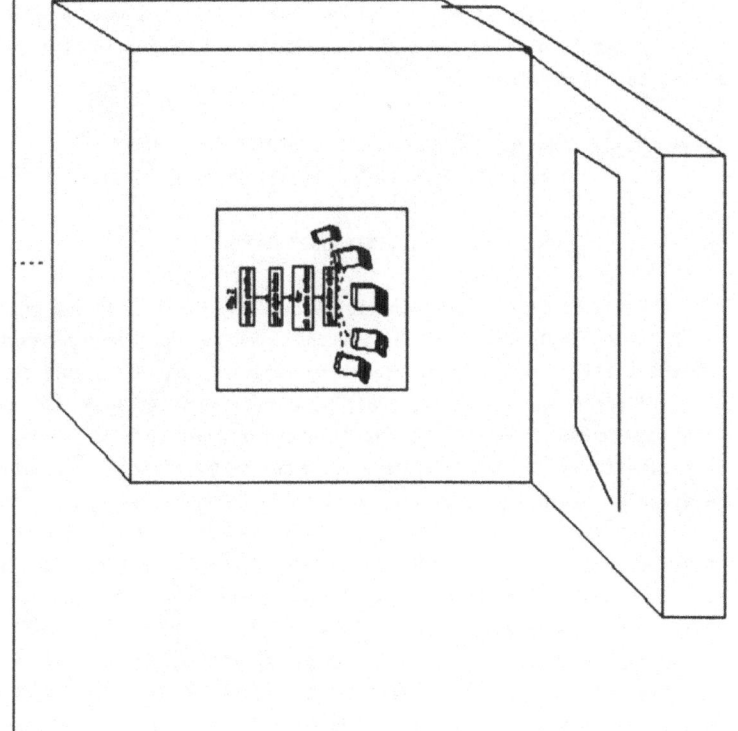

AOV : THE PRISM MODULE FOR ANALYSING DESIGNED EXPERIMENTS

R.W.PAYNE

Statistics Department, Rothamsted Experimental Station,
Harpenden, Herts

SUMMARY

A brief description is given of the scope and syntax of AOV,
together with an example illustrating the sort of output that
can be obtained.

Keywords: ANALYSIS OF VARIANCE;   COMPUTER PROGRAMS;
            STATISTICAL COMPUTING;   STATISTICAL PACKAGES

1. INTRODUCTION

The AOV module of Prism was initially conceived as a sister program to
GLIM, which would be formed by linking the GLIM housekeeping routines
to the analysis of variance algorithm from Genstat Mk. 3.09. Package
comparisons (e.g. Heiberger, 1981) have shown Genstat ANOVA to contain
the most complete facilities available for the analysis of designed
experiments, so it was thought worthwhile to make these accessible to
a wider class of users, particularly those on smaller computers.

The initial work (by J.D.Beasley and myself of Rothamsted, and
B.P.Murphy and J.D.Henstridge of the University of Western Australia)
was intended to be linked to GLIM 3. However it soon became apparent
that preparations for GLIM 4 were well under way and that the
housekeeping routines (now known as the Kernel) could be written with
AOV, as well as GLIM, in mind - thus greatly simplifying our task.
Consequently the program was reframed as a module of Prism (as the
whole package is now known). Improvements in Mk. 4.02 Genstat ANOVA
made it worthwhile starting again from the Genstat code, and this time
care has been taken to note, by comment cards, the differences between
the source code of the two packages. Future improvements can thus be
made in parallel in both programs. I gratefully acknowledge the help
of R.P.White (of Rothamsted) and J.D.Henstridge with some of the new
subroutines and changes to the code. R.J.Baker's assistance with the

parts of the Kernel pertaining to AOV has also been invaluable.

2. THE SCOPE OF AOV

To the GLIM purist, AOV may seem rather limited - analysis of variance is, after all, merely a generalised linear model with Normal errors and identity link. However, AOV differs from GLIM in two important ways. Firstly, it can handle more than one error term. Secondly, it provides a very much more comprehensive output, as shown in the example below.

AOV uses Wilkinson's (1970) algorithm (see also Payne and Wilkinson, 1977) and the designs that it can handle are termed balanced. These include all orthogonal designs (e.g. single-factor designs, equally or proportionally replicated factorials, randomised blocks, Latin and Graeco-Latin squares, split plots etc), designs with balanced confounding (e.g. balanced incomplete block designs, balanced lattices), as well as some designs with partially balanced confounding. To indicate the scope of the algorithm, a data file has been prepared showing how to analyse all the worked examples from Cochran and Cox (1957). Similar files will be prepared showing examples from non-agricultural applications.

3. SYNTAX

To enable AOV to deal with multiple error terms, two formulae are required to specify the analysis (as opposed to the single directive $FIT in GLIM). The structure of the design - and thus the error terms for the analysis - are specified by $BLOCKS. In the example in section 4,

$BLOCKS BLOC / PLOT / SUBP

defines a split-plot design with sub-plots (SUBP) nested within whole plots (PLOT) which are nested within blocks (BLOC). There are three error terms: BLOC, representing the random variation of the blocks; BLOC.PLOT for differences between whole plots within blocks; and BLOC.PLOT.SUBP for sub-plots within whole plots. For less complicated designs, the formula is simpler - a randomised blocks design just has plots nested within blocks and would be specified by

$BLOCKS BLOC / PLOT .

Latin squares have rows crossed with columns: i.e.

$BLOCKS ROWS * COLS .

Other examples will be given in the AOV User's Guide. If there is a single error term (as with completely randomised designs) $BLOCKS can be omitted.

The systematic aspects of the design - those that are imposed by the experimenter - are specified by $TREATMENTS. In the example there are two treatment factors: variety (V) which was randomised onto the whole plots within each block, and nitrogen (N) which was applied to sub-plots. To estimate the main effects of V and N and the interaction V.N, the formula would be

        $TREATMENTS V * N .

In the example the nitrogen sum of squares has been partitioned, by orthogonal polynomials, into linear, quadratic and cubic effects. For this the specification is

        $TREATMENTS V * N<3>

(where <3> indicates polynomial contrasts to order 3).

Any covariates that are required in the analysis are listed following the directive $COVARIATES.

Once the design has been specified, the analysis of a variate Y, say, can be performed by

        $ANOVA Y .

By default this produces the analysis of variance table (see example).

Further output can be obtained by $EXHIBIT, which parallels the directive $DISPLAY in GLIM (although, for obvious reasons, the syntax is slightly different). The output components and corresponding keywords are:

AOV   -   analysis of variance table;
INFO   -   information summary (details of non-orthogonality, aliasing etc);
COVA   -   covariate regression coefficients, with their standard errors etc.;
EFFE   -   estimated effects for treatment terms;
SUBM   -   estimated coefficients and standard errors for submodel contrasts (orthogonal polynomials etc);
RESI   -   residuals;
MEAN   -   tables of means for treatment terms,
              which may be accompanied by either

SED   -   standard errors for differences between pairs of means,

                   or

SEM   -   standard errors of means;

CV   -   coefficients of variation, residual degrees of freedom and plot standard errors for each stratum (or term in the BLOCK formula);

MV   -   estimated missing values.

Many of these are illustrated in the example. The order above indicates both the order in which components are printed if several are specified in a single $EXHIBIT and the precedence for determining which component is printed if a keyword is abbreviated to less than 4 letters. For example $EXHIBIT C prints covariate regression coefficients, as COVA comes before CV in the list.

There is also a directive $DEXHIBIT, with similar syntax to $EXHIBIT, which controls the default output produced by $ANOVA. This will be particularly useful when running Prism in batch mode, when there are no longer constraints such as those arising from the size of the user's video screen, or the printing speed of a teletype. It may then be sensible to request all the required output by default. For example, $DEXHIBIT AOV MEANS SEM ensures that subsequent $ANOVA directives will produce tables of means and standard errors, as well as the analysis of variance table.

Other directives in the AOV module are described briefly below.

$ALG   -   enables the user to modify parameters and tolerances of the AOV algorithm.

$OBT   -   (OBTAIN) enables output components from an analysis of variance to be copied into standard Prism structures named by the user. This will allow macros to be written for combination of information in partially confounded designs etc.

$LIM   -   defines limits governing the inclusion of model terms in the analysis. (For example, $LIMITS FACT 3 specifies that factorial terms containing up to 3 factors will be included.)

$WAV   -   specifies the variate of weights for weighted analysis of variance.

These will all be described in more detail in the AOV User's Guide.

4. EXAMPLE

The output below shows the transcript file from an interactive run in which a split-plot design was analysed. The design (which has also been used in the past to illustrate Genstat ANOVA) is taken from Yates (1937) and is a varietal trial of oats. There are two treatment factors, V, representing the 3 varieties of oats: ("Victory", "Golden Rain II" and "Marvellous") and N, corresponding to 4 levels of nitrogen fertiliser (0, 0.2, 0.4 and 0.6 cwt manure per acre). A split-plot design was chosen because it is more convenient to work with larger plots for varieties than for fertilisers. The design was thus set up in two stages. In the first the 6 blocks were each divided into 3 whole plots and the 3 varieties were allocated at random to the plots within each block. Then each whole plot was divided into 4 subplots, onto which the 4 fertilisers were randomised.

Notice that, in the analysis of variance table, the program has automatically determined the error term against which each treatment should be compared. - Varieties, which were applied to whole plots, are tested against the residual term for whole plots, while nitrogen (applied to sub plots) and variety.nitrogen are tested against the sub-plot residual. The sum of squares for nitrogen is partitioned into three terms, "(1)", "(2)" and "(3)", representing linear, quadratic and cubic effects of nitrogen. Similarly, under V.N, the term "DEV.(1)" represents the effect of fitting a different linear coefficient for each variety. Knowledge of the appropriate error term for each treatment term also enables the correct standard errors to be calculated for the tables of means, effects etc. Apart from the use of perhaps unfamiliar terms like "stratum" (which will be explained in more detail in the AOV User's Guide) the output takes a fairly standard form (see, for example, Cochran and Cox, 1957, or John and Quenouille, 1977).

To make the output below easier to follow, the directives that have been typed in from the terminal have been underlined. This has been done by the author not by Prism. The syntax and output are as of 1st June 1982. There may be slight modifications in the final release version.

PRISM-1.01 (C)1982 ROYAL STATISTICAL SOCIETY, LONDON

```
$ECHO ON $INPUT 16 80 $
$UNITS 72
$FACTOR N / (CW.0,CW.2,CW.4,CW.6)
     : V / (VICT,GOLD,MARV)
     : BLOC / 6 ~ 12
     : PLOT / 3 ~ 4
     : SUBP / 4 ~ 1
$DATA * N $READ
_CW.6 _CW.4 _CW.2 _CW.0 _CW.0 _CW.2 _CW.6 _CW.4 _CW.0 _CW.2 _CW.4 _CW.6
_CW.4 _CW.0 _CW.2 _CW.6 _CW.6 _CW.0 _CW.2 _CW.4 _CW.2 _CW.0 _CW.4 _CW.6
_CW.2 _CW.4 _CW.6 _CW.0 _CW.6 _CW.2 _CW.4 _CW.0 _CW.0 _CW.6 _CW.2 _CW.4
_CW.4 _CW.6 _CW.0 _CW.2 _CW.0 _CW.4 _CW.6 _CW.2 _CW.2 _CW.4 _CW.6 _CW.0
_CW.6 _CW.0 _CW.4 _CW.2 _CW.4 _CW.6 _CW.0 _CW.2 _CW.4 _CW.6 _CW.2 _CW.0
_CW.4 _CW.0 _CW.6 _CW.2 _CW.6 _CW.4 _CW.0 _CW.2 _CW.0 _CW.2 _CW.4 _CW.6
$DATA * V $READ
_MARV _MARV _MARV _MARV _VICT _VICT _VICT _VICT _GOLD _GOLD _GOLD _GOLD
_MARV _MARV _MARV _MARV _VICT _VICT _VICT _VICT _GOLD _GOLD _GOLD _GOLD
_GOLD _GOLD _GOLD _GOLD _MARV _MARV _MARV _MARV _VICT _VICT _VICT _VICT
_MARV _MARV _MARV _MARV _GOLD _GOLD _GOLD _GOLD _VICT _VICT _VICT _VICT
_GOLD _GOLD _GOLD _GOLD _VICT _VICT _VICT _VICT _MARV _MARV _MARV _MARV
_VICT _VICT _VICT _VICT _GOLD _GOLD _GOLD _GOLD _MARV _MARV _MARV _MARV
$DATA * Y $READ
156 118 140 105 111 130 174 157 117 114 161 141
104 70 89 117 122 74 89 81 103 64 132 133
108 126 149 70 144 124 121 96 61 100 91 97
109 99 63 70 80 94 126 82 90 100 116 62
96 60 89 102 112 86 68 64 132 124 129 89
118 53 113 74 104 86 89 82 97 99 119 121
$BLOCKS BLOC/PLOT/SUBP
$TREATMENTS V*N<3>
$RETURN
$ANOVA Y
```

```
---- AOV  SOURCE      DF        SS        MS     VR

BLOC STRATUM          5     15875.3    3175.1

BLOC.PLOT STRATUM
V                     2      1786.4     893.2    1.49
RESIDUAL             10      6013.3     601.3

BLOC.PLOT.SUBP STRATUM
N                     3     20020.5    6673.5   37.69
   (1)                1     19536.4   19536.4  110.32
   (2)                1       480.5     480.5    2.71
   (3)                1         3.6       3.6    0.02
V.N                   6       321.7      53.6    0.30
  DEV.(1)             2       168.3      84.2    0.48
  DEV.(2)             2        11.1       5.5    0.03
  DEV.(3)             2       142.3      71.2    0.40
RESIDUAL             45      7968.7     177.1

TOTAL                71     51985.9
```

$EXHIBIT MEAN SED :

```
---- TABLES OF MEANS

   GRAND MEAN      104.0

   V      VICT     GOLD     MARV
          97.6    104.5    109.8

   N      CW.0     CW.2     CW.4     CW.6
          79.4     98.9    114.2    123.4

   V    N    CW.0      CW.2      CW.4      CW.6
VICT          71.5      89.7     110.8     118.5
GOLD          80.0      98.5     114.7     124.8
MARV          86.7     108.5     117.2     126.8
```

--- STANDARD ERRORS OF DIFFERENCES OF MEANS

TABLE	V	N	V
			N
REP	24.	18.	6.
SED	7.08	4.44	9.72

EXCEPT WHEN COMPARING MEANS WITH SAME LEVEL(S) OF

V			7.68

EFFECTS :

--- BLOC.PLOT STRATUM

V EFFECTS   ESE   5.0 REP   24.

V	VICT	GOLD	MARV
	-6.3	0.5	5.8

--- BLOC.PLOT.SUBP STRATUM

N EFFECTS   ESE   3.1 REP   18.

N	CW.0	CW.2	CW.4	CW.6
	-24.6	-5.1	10.2	19.4

V.N EFFECTS   ESE   5.4 REP   6.

V	N	CW.0	CW.2	CW.4	CW.6
VICT		-1.5	-2.9	3.0	1.5
GOLD		0.1	-0.9	-0.1	0.9
MARV		1.5	3.8	-2.9	-2.4

SUBMODELS :

--- BLOC.PLOT.SUBP STRATUM

-- N CONTRASTS

```
(1)   14.7  SE      1.4  SS DIV    90.0
(2)   -2.6  SE      1.6  SS DIV    72.0
(3)   -0.3  SE      2.3  SS DIV    32.4
```

-- V.N CONTRASTS

DEV.(1) ESE   2.4 SS DIV   30.0

```
    V      VICT    GOLD    MARV
           1.5     0.3    -1.8
```

DEV.(2) ESE   2.7 SS DIV   24.0

```
    V      VICT    GOLD    MARV
          -0.0     0.5    -0.5
```

DEV.(3) ESE   4.0 SS DIV   10.8

```
    V      VICT    GOLD    MARV
          -2.4    -0.3     2.7
```

RESIDUALS :

BLOC RESIDUALS   SE   14.8 REP   12.

```
  BLOC       1        2        3        4        5        6
            31.4     -5.8      3.3    -13.1     -8.1     -7.7
```

BLOC.PLOT RESIDUALS   SE   9.1 REP   4.

```
BLOC PLOT       1        2        3
   1          -11.4     14.0     -2.6
   2           -9.0     -0.3      9.3
   3            5.5      8.2    -13.7
   4          -11.5      4.1      7.4
   5           -9.7     -7.1     16.8
   6           -0.4     -6.5      6.9
```

```
BLOC.PLOT.SUBP  RESIDUALS   SE   10.5  REP      1.
```

BLOC	PLOT	SUBP	1	2	3	4
1	1		9.2	-19.1	11.5	-1.6
	2		-5.9	-5.0	10.1	0.8
	3		8.2	-13.3	17.6	-12.6
2	1		1.6	-1.9	-4.7	5.0
	2		9.6	8.6	5.5	-23.7
	3		1.0	-19.5	13.8	4.7
3	1		0.7	2.6	15.4	-18.7
	2		5.7	4.0	-7.6	-2.1
	3		-0.1	-8.1	11.7	-3.5
4	1		16.4	-3.3	0.9	-14.0
	2		9.0	-11.7	10.2	-7.5
	3		6.0	-5.2	3.1	-3.9
5	1		-11.1	-2.3	-7.9	21.2
	2		16.3	-17.4	11.6	-10.5
	3		6.1	-11.5	11.8	-6.4
6	1		15.3	-10.4	2.6	-7.5
	2		-6.6	-14.4	23.2	-2.3
	3		11.1	-8.7	2.6	-5.0

CV :

---- STRATUM	DF	SE	CV%
BLOC	5	16.27	15.6
BLOC.PLOT	10	12.26	11.8
BLOC.PLOT.SUBP	45	13.31	12.8

MV :

---- NO MISSING VALUES

INFORMATION $

---- INFORMATION SUMMARY
ALL TERMS ORTHOGONAL, NONE ALIASED

$STOP

5. CONCLUSION

It is hoped that the AOV module will provide a significant enhancement to Prism. The example above was chosen to exhibit the power of the program without choosing too complicated a design. Further examples - both simpler and more complicated - will be given in the User's Guide.

REFERENCES

COCHRAN, W.G. and COX, G.M. (1957). Experimental Designs, 2nd Edition. New York: Wiley.

HEIBERGER, R.M. (1981). The Specification of Experimental Designs to Anova Programs. American Statistician, 35, 98-108.

JOHN, J.A. and QUENOUILLE, M.H. (1977). Experiments: Design and Analysis, 2nd Edition. London: Charles Griffin.

PAYNE, R.W. and WILKINSON, G.N. (1977). A General Algorithm for Analysis of Variance. Applied Statistics, 26, 251-260.

WILKINSON, G.N. (1970). A General Recursive Algorithm for Analysis of Variance. Biometrika, 57, 19-46.

YATES, F. (1937). The Design and Analysis of Factorial Experiments. Commonwealth Bureau of Soil Science, Technical Communication No.35.

<u>THE   APL   ALTERNATIVE</u>

P.J.Green

Department of Mathematics
University of Durham

Summary

APL is an interactive computing language.   In this paper we discuss its
suitability for statistical analysis and experimentation, and in particular for
maximum likelihood estimation using iteratively reweighted least squares.

Keywords

APL;   interactive statistics;   generalised linear models;   iteratively
reweighted least squares;   resistant regression;   GLIM.

1. AN INTRODUCTION TO APL

APL is a programming language developed from the ideas in K.E.Iverson's book
(1962) which described a new notation for the concise expression of mathematical
algorithms.   Both in its initial implementation for the IBM 360 series, and in its
later adoption by most mainframe manufacturers, the emphasis has been on interactive
use.   As with most high-level languages, dialects were developed for different machine ranges, with subsequent belated attempts at standardisation.   Nevertheless the
language is remarkably standard, with most differences lying in the communication
between APL and the resident operating system.

Recent developments in availability have followed from considerable inroads into
the business computing market following the adoption of APL by systems bureaux.
Powerful microcomputer implementations of the language are now available which have
encouraged the introduction of APL into small commercial installations.   Nevertheless
scientific and especially university use of APL is still likely to be from a terminal
connected to a large mainframe.

Some of the features of APL that are relevant to this paper are:
(a) Data structures are arrays (scalars, vectors, matrices, and of higher dimension)
    of type numerical or literal.
(b) There are no data declarations, and storage is allocated dynamically.

(c) Subprograms, which are called functions, are used where other languages use operators, so their results are immediately available as arguments to other functions.

(d) Arithmetic and mathematical functions apply component-wise to arrays.

(e) Matrix handling includes selection and indexing, generalised transpose, catenation and lamination.

(f) Generalised inner and outer products, least-squares solution of linear equations, sorting and random number generation are all built in.

(g) Input and output are free-format by default.

(h) Variables and functions are stored together in a workspace (in fact a file on disc, read into memory for manipulation and execution).

APL is extremely easy to learn at least for the mathematically minded as all these features are available in a simple syntax. Unlike those of most other languages, the rules do not depend on context. One example is that there are no operator priorities: execution always proceeds from right to left. In most implementations, APL is interpreted not compiled, but when the basic data item is the array not the scalar, looping is often avoided, and the interpretive burden is acceptable.

One obvious feature of APL not mentioned so far is the unorthodox character set. Part of the compactness of APL notation is derived from the use of a rich variety of special symbols to denote system functions. However, this unfamiliar symbolism provokes an initial inhibition against learning the language, necessitates expensive special purpose terminals, and must be considered a mistake. Attempts to circumvent these disadvantages include D.A.Evans' APL84 in which many symbols are replaced by keywords (Evans, 1980), and a portable interpreter for a new language similar to APL under development by the author; both of these systems use standard ASCII terminals.

It is possible to write extremely compact APL programs that perform complicated calculations. Although there is some advantage in machine efficiency in doing so (partly because the naming of intermediate results can then often be avoided, and partly because interpretation is then faster), this practice should be discouraged. It has led to an impression that APL is a "write-only" language in which programs are written but never read, and has discouraged some potential users.

However there is some validity in the impression that APL is for those who wish to "get their hands dirty" in computing. The great versatility of the language, which means it can be sensible to write a program that will be used only once, is paid for by comparatively less protection and explanation offered to the naive user. There have been attempts to customise and protect by producing packages within APL in which the user does not actually communicate in the language, but APL should not be judged by these. It is above all a language in which one writes solutions to problems, not one for producing packages.

2. APL IN STATISTICS

How is all this related to the requirements of statistical computing? Let us
neglect the large-scale problems of data-reduction that often arise from sample surveys, certain types of routine analysis justifying special purpose programs, and all
attempts to reduce the computational side of statistical practice to a "black box".
Then we require interactive access, allowing manipulation of data and the ability to
explore as well as to follow prescribed steps of analysis.

It should be clear that even without the addition of specifically statistical
functions, APL incorporates a large number of the ingredients for this sort of statistical programming. With data arrays and derived statistics represented as APL
variables, the language offers a flexibility in data-handling afforded by no available
packages. Variates and qualitative factors are easily handled, selection and sorting
are immediately available, and multiple regression built in at the machine code level.
Unusual non-standard data formats can be decoded using APL's string-handling capabilities. Furthermore, this flexibility is obtained in a compact and attractive notation, unencumbered by the redundancies of directive-driven packages.

User-defined functions provide the next building block, usually performing small
stages of analysis, and either recalled from some earlier project, or especially written for the purpose. As familiarity with a data set is gained, such functions can
be combined into higher-level functions implementing a complete analysis. Global
variables allow the implicit storing of intermediate results, and of output structures
that may not always be of interest. And the workspace concept permits data and
methods to be stored together.

APL was not specifically designed for statistical work, but obviously has highly
relevant features, and is receiving an increasing amount of attention (see, for
example, Anscombe(1981), Evans (1981) and Besag and Green (1981)). It is interesting to note that PRISM has recognised the usefulness of the APL array handling features by adopting some of them in the ACAL directive. While APL is more suited to
some problems than others, it is a general-purpose language and so capable of dealing
with most statistical computing of the type we have described. What we can perhaps
expect is that when faced with a high-quality interactive package specialised to deal
with a particular type of problem (e.g. GLIM), it will not be competitive. This must
be regarded as a comparison on terms unfair to APL, yet I hope to show in the next
sections that the contest is not one-sided.

3. APL AND GENERALISED LINEAR MODELS

Generalised linear models as defined by Nelder and Wedderburn (1972) form a
class of distributions from a one-parameter exponential family, with optional nuis-

ance scale parameter.   They are of interest statistically because they include several standard models.   Computationally they are unified by the simple algorithm that was suggested by Nelder and Wedderburn for the iterative solution for the maximum likelihood estimates.   The Newton-Raphson procedure, using expected second derivatives, reduces to a least-squares solution for each update of the estimate values. Thus we call the method IRLS (iteratively reweighted least squares).   It is a generalisation of that used by Finney (1947) for probit analysis, in which at each step we recalculate a working "dependent variate" and a weighting variate and perform a weighted least squares regression.

GLIM is basically a sophisticated IRLS algorithm, providing several standard models automatically and a degree of monitoring and control of convergence, with a package of associated data-handling directives.

An alternative approach is to take a language providing all the data-handling capabilities one might require, and add an IRLS algorithm to it.   This is possible in APL because the computations in the IRLS step consist of component-wise arithmetic, matrix-product, and least squares solution, plus some mathematical functions in specifying the model, and these are all provided in the APL system.

Lest this be considered re-inventing the wheel, one can answer firstly that perhaps GLIM did that, and secondly that the labour of re-invention is so slight that it is paid for by the resulting flexibility of APL.

There are of course disadvantages in such an approach.   The main one is lack of control over the least squares function in APL.   Only one computational method is provided, the whole design matrix must be stored, and no updating of a fit by adding extra covariates is possible.   Further, while the usual ambiguities in factorial design notation necessitating side conditions on parameters are easily handled (by specifying columns of the design matrix appropriately), "accidental" aliasing will be detected but not located and isolated as in GLIM.

4. EXTENDING GENERALISED LINEAR MODELS

GLIM and APL versions of IRLS both have capabilities beyond that of the "standard" models.

We are dealing with a log-likelihood function $L(\underline{\eta};\underline{y})$ given an n-vector of observations $\underline{y}$.   The parameter $\underline{\eta}$, usually also an n-vector, is a function of the parameters of interest, a p-vector $\underline{\beta}$: $\underline{\eta} = \underline{\eta}(\underline{\beta})$.

If we write $\underline{u}$ for the vector $\partial L/\partial \underline{\eta}$, and D for the $n \times p$ Jacobian matrix $\partial \underline{\eta}/\partial \underline{\beta}$

then the likelihood equations are $D^T\underline{u} = 0$. For Newton-Raphson iteration we need the second derivatives of L with respect to $\underline{\beta}$. In IRLS it is usual to approximate these by the matrix $-D^T AD$ where A is the $n \times n$ matrix $E(-\partial^2 L/\partial \underline{n}^2)$. For a generalised linear model this is just the so-called Fisher scoring method - we are using expected second derivatives; for other models a second linearising approximation has been made. Thus the IRLS step is to update the parameter $\underline{\beta}$ to $\underline{\beta}^*$ where

$$(D^T AD)(\underline{\beta}^* - \underline{\beta}) = D^T \underline{u}$$

and $\underline{u}$, D and A are calculated at the current value $\underline{\beta}$. As equations in $\underline{\beta}^*$, these have the form of normal equations for a regression of $(A^{-1}\underline{u} + D\underline{\beta})$ on the columns of D with weight matrix A. The resulting regression coefficients form the updated $\underline{\beta}^*$; from these are calculated new values of $\underline{u}$, D and A, and the cycle repeated until convergence.

Notice that we have concentrated on solving the likelihood equations: if the process converges we are guaranteed a solution. Separate reassurance may be needed that this point maximises the likelihood, if we are working with an unfamiliar model.

Generalised linear models as implemented in GLIM concern a special case of the above, in three respects:

(a) Because of the exponential family assumption, $\underline{u}$ is simply specified in terms of $\underline{y}$, its mean and variance, and the "link function".

(b) Because the observations must be independent, A is diagonal, so the weighted least squares step is facilitated.

(c) Because the relationship between $\underline{n}$ and $\underline{\beta}$ is linear, $\underline{n} = X\underline{\beta}$ and so D is constant, the design matrix X.

For standard models these ingredients are provided automatically, and the user merely selects an appropriate X. The macro facility permits certain other models to be specified. System variates such as %FV, %VA and %DR must be assigned values so that GLIM will calculate the correct $\underline{u}$. Much ingenuity has been expended on forcing problems into this mould.

Such ingenuity is misdirected since the resulting macros may easily consist of more complicated coding than an entire APL function purpose-built for the problem.

Using macros in GLIM, requirement (a) above can be relaxed considerably. It is a nuisance to relax (c) and apparently impossible to do anything about (b). Models requiring a non-diagonal "information matrix" A are currently receiving much attention, as also is the possibility of using observed rather than expected information (see Jørgensen (1982) and McCullagh (1982)). These are easily dealt with in APL using a function for Cholesky decomposition.

An APL implementation of IRLS permits generalisation in other directions. Firstly, we have not so far raised the question of dealing with nuisance parameters. GLIM permits a common unknown scale nuisance parameter, and provides facilities for controlling it. But in generalised linear models, this scale parameter factors out

of the problem.   In more general estimation problems, for example linear regression
with a specified but non-Gaussian error distribution, the scale parameter must be
either held fixed, or simultaneously iterated with the structural parameter $\underline{\beta}$ of main
interest.   This is related computationally, but not conceptually, to the matter of
resistant alternatives to maximum likelihood, that is, to solutions of weighted likelihood equations:

$$\sum_i w(z_i) \frac{\partial L_i}{\partial \beta_j} = 0 \quad \text{for all } j.$$

Here L_i is the contribution to the log-likelihood and z_i an appropriately defined
residual for the ith observation, and $w(\ )$ a weighting function chosen to down-
weight the influence of discrepant observations.   Such fits have been discussed by
Pregibon (1982) who has GLIM macros for the purpose.   But again, this problem is
more conveniently handled in APL.

A further useful possibility is that of alternative initial values for the
iteration: in APL one can for example provide an L_1 (least absolute deviation) solution as an option.

One conclusion from this is that it would be practicable to write a universal
IRLS package in APL.   But that is not the point I wish to make.   It is also feasible
to write a purpose-built analysis for each problem encountered.

5. OTHER APPLICATIONS OF APL

Generalised linear and related models are well suited to APL because of the
matrix and regression capabilities we have discussed.   But of course the APL approach
is by no means dedicated to such models.   The consulting statistician with APL at
his fingertips is immediately prepared to deal with his next customer, who is not
interested in logistic regression, but in EDA or time series analysis or contingency
tables or concordance.   An idea of the scope of potential applications can be gained
from the recent book by Anscombe (1981).

APL has also been used in teaching statistics.   The statistics and medical
statistics departments at the University of Newcastle, for example, have found that
APL allows students more time for interpretation by saving labour on arithmetic
(Evans, 1981).

APL has its peculiarities and its disadvantages.   It is not well suited to
large data-sets or to much routine analysis.   It does require familiarity and some
confidence in the user to be effective.   But it is an outstandingly useful tool for
the statistician to have at hand.

"Purpose-built" programming frees the statistician and the problem to choose the
analysis, and avoids the package moulding the problem.

ACKNOWLEDGEMENTS

I would like to thank Julian Besag, who first told me about APL, and with whom
I have had many discussions about statistical computing.

REFERENCES

Anscombe, F. (1981) Computing in statistical science through APL. Springer-Verlag,
    New York.
Besag, J.E. & Green, P.J. (1981) APL in statistical research. Unpublished manuscript
    of paper presented to RSS General applications section, December 1981.
Evans, D.A. (1980) APL84 - an interactive APL-based statistical computing package.
    In COMPSTAT 1980, 241-245. Physica-Verlag, Vienna.
Evans, D.A. (1981) Some basic features of APL. Unpublished manuscript of paper
    presented to RSS General applications section, December 1981.
Finney, D.J. (1947) Probit analysis. Cambridge University Press.
Iverson, K.E. (1962) A programming language. Wiley, New York.
Jørgensen, B. (1982) Maximum likelihood estimation and large sample inference for
    generalised linear and non-linear regression models.   To appear.
McCullagh, P. (1982) Quasi-likelihood functions.   To appear.
Nelder, J.A. & Wedderburn, R.W.M. (1972) Generalized linear models. J.R.Statist.
    Soc. A, 135, 370-384.
Pregibon, D. (1982) Resistant fits for some commonly used logistic models with
    medical applications.   Biometrics, 38, 485-498.

DIRECT LIKELIHOOD INFERENCE

Murray Aitkin
Centre for Applied Statistics
University of Lancaster

SUMMARY

A unified approach to statistical modelling is proposed through direct likelihood inference. An example is given for the choice of a regression model for a complex cross-classification.

KEYWORDS: LIKELIHOOD, CONFIDENCE INTERVAL, STATISTICAL MODELS, INFERENCE

1. INTRODUCTION

Successive releases of GLIM have provided us with increasingly powerful tools for the fitting of models to data by maximum likelihood. PRISM, with its new GRAPH module, extends our abilities even further, particularly in the direction of plotting of response surfaces and several-dimensional functions. The macro facility and the ability to fit one's own model allow a wide range of non-standard problems to be incorporated into the general linear model framework, using for example the EM algorithm or a composite link function.

We are now, I believe, at a point where a unified approach to statistical analysis is possible. While maximum likelihood estimation, and likelihood ratio testing, are in widespread use, they are by no means universally accepted as a standard framework for statistical inference, even by non-Bayesians. The small-sample bias of maximum likelihood estimators, and their sometimes poor mean square error properties when compared with other estimators, have led to a search for better estimators, and many computer years of simulation studies.

I believe that this search for better estimators, the evaluation of such estimators by their bias and mean square error, and indeed the whole concept of a point estimator, have been largely irrelevant to the real problems of statistics, and have led to a major diversion of effort into unfruitful paths.

I will not attempt in this paper to justify or discuss this view. What I will do is to illustrate what I believe to be a unified approach to analysis based on direct likelihood. This approach is not original : it was proposed first by Fisher (1912) and has been developed extensively by Barnard (1949, 1962), Sprott and Kalbfleisch (1969, 1970), Edwards (1972) and many others.

What is perhaps new in the present discussion is the confidence interpretation of likelihood regions, and the insight this gives into the contentious area of regression model choice in complex cross-classified data, which I will illustrate later. Let us begin with some simple examples.

2. THE NORMAL DISTRIBUTION AND NORMAL LIKELIHOODS

Consider a large population, on each member of which we are able to measure a response y with high precision: a measured value y corresponds to a "true" value in the interval $y \pm \delta/2$. Suppose the approximating population model for y is a normal distribution $N(\mu, \sigma^2)$: that is, the probability $P(y)$ of drawing a single observation y is given approximately by

$$P(y) = \phi(\frac{y + \delta/2 - \mu}{\sigma}) - \phi(\frac{y - \delta/2 - \mu}{\sigma})$$

$$\approx \frac{\delta}{\sigma} \phi(\frac{y - \mu}{\sigma})$$

$$= \frac{\delta}{\sigma\sqrt{2\pi}} \exp (-\tfrac{1}{2} \frac{(y - \mu)^2}{\sigma^2})$$

since the measurement precision is high (δ small).

Given a random sample $y_1, \ldots, y_n$ drawn with replacement from the population, the likelihood function is proportional to

$$P(y_1, \ldots, y_n) = \frac{\delta^n}{\sigma^n (2\pi)^{n/2}} \exp (-\tfrac{1}{2} \frac{\Sigma(y_i - \mu)^2}{\sigma^2})$$

Let us consider several cases.

2.1 σ known

The likelihood function can be taken to be

$$L(\mu) = \exp (-\tfrac{1}{2} \frac{\Sigma(y_i - \mu)^2}{\sigma^2})$$

with log-likelihood function

$$\ell(\mu) = -\tfrac{1}{2} \frac{\Sigma(y_i - \mu)^2}{\sigma^2}$$

Following through the usual maximum likelihood calculations, we have

$$\ell'(\mu) = \Sigma(y_i - \mu)/\sigma^2$$

$$\ell''(\mu) = -n/\sigma^2$$

and all higher derivatives are zero: the log-likelihood is quadratic in μ. The ML estimator of μ is $\bar{y}$, and from large-sample maximum likelihood theory we know that $(\hat{\mu} - \mu) I_{\hat{\mu}}^{\frac{1}{2}}$ is asymptotically $N(0,1)$, where $I_{\hat{\mu}} = -\ell''(\hat{\mu})$. The result in this case is exact, since $\bar{y} \sim N(\mu, \sigma^2/n)$. An asymptotic $100(1-\alpha)\%$ confidence interval for μ (again, exact in this case) is obtained from $\hat{\mu} \pm \lambda_{\alpha/2} I_{\hat{\mu}}^{-\frac{1}{2}}$, or $\bar{y} \pm \lambda_{\alpha/2} \sigma/\sqrt{n}$.

The same results can be obtained from the likelihood function by a direct likelihood approach. We first define the relative likelihood function:

$$R(\mu) = L(\mu)/L(\hat{\mu})$$

$$= \exp (-\frac{1}{2\sigma^2} [\Sigma(y_i - \mu)^2 - \Sigma(y_i - \bar{y})^2])$$

$$= \exp\left(- \frac{n(\bar{y}-\mu)^2}{2\sigma^2}\right)$$

In likelihood theory, $R(\mu)$ gives the plausibility of, or support for, different values of μ, with reference to the most plausible or best supported value $\hat{\mu} = \bar{y}$. A $100\gamma\%$ likelihood interval for μ consists of those values of μ whose relative likelihoods are at least γ: $R(\mu) \geq \gamma$. This inequality is equivalent to

$$\frac{n(\bar{y} - \mu)^2}{\sigma^2} \leq -2 \log \gamma = c_\gamma^2$$

or

$$\mu \in (\bar{y} \pm c_\gamma \sigma/\sqrt{n})$$

This is the same confidence interval as above, if we identify c_γ with $\lambda_{\alpha/2}$. That is, a $100\gamma\%$ likelihood interval for μ has a confidence coefficient in repeated sampling given by

$$\chi^2_{1,\alpha} = \lambda^2_{\alpha/2} = -2 \log \gamma \quad , \quad \gamma = \exp(-\tfrac{1}{2}\chi^2_{1,\alpha})$$

Thus for example if $\gamma = .146$, $\alpha = .95$: a 14.6% likelihood interval has a confidence coefficient of 95% in repeated sampling. If $\gamma = .036$, $\alpha = .99$, etc.

An important part of likelihood inference is plotting the (relative) likelihood function. We can see directly that $R(\mu)$ has the form of a normal density function in μ: equivalently, $\log R(\mu)$ is quadratic in μ. Further, there is a symmetry in $\bar{y}$ and μ in $R(\mu)$: the relative likelihood is also, apart from the constant $\frac{\sqrt{n}}{\sqrt{2\pi}\sigma}$, the normal density function of $\bar{y}$ in repeated sampling. A moment's thought shows why this is so: to obtain the distribution of $\bar{y}$ in random sampling, we write down the joint density of $y_1,\ldots,y_n$, (which is the same as the likelihood function), make a Helmert transformation to $\bar{y}$ and $(n-1)$ functions of differences, and integrate or factor out the $(n-1)$ other variables. Since $\bar{y}$ is sufficient for μ, we are left with a multiple of the likelihood function as the density function of $\bar{y}$, this multiple being independent of μ.

We are thus often able to obtain immediately the sampling distribution of the ML estimator from the likelihood function as Fisher (1934) pointed out. In particular, a normal relative likelihood for θ implies a normal distribution for $(\hat{\theta}-\theta)I_{\hat{\theta}}^{\frac{1}{2}}$. The importance of this result will be seen in the next example.

2.2 μ known

The likelihood function can be taken as

$$L(\sigma) = \frac{1}{\sigma^n} \exp(-\tfrac{1}{2}\frac{\Sigma(y_i-\mu)^2}{\sigma^2})$$

It is conventional to think of σ^2 as the parameter of interest. If we put $\theta = \sigma^2$ then

$$L(\theta) = \frac{1}{\theta^{n/2}} \exp \left(-\tfrac{1}{2} \frac{\Sigma(y_i-\mu)^2}{\theta}\right)$$

$$\ell(\theta) = -\frac{n}{2} \log\theta - \tfrac{1}{2}\frac{\Sigma(y_i-\mu)^2}{\theta}$$

$$\ell'(\theta) = -\frac{n}{2\theta} + \tfrac{1}{2}\frac{\Sigma(y_i-\mu)^2}{\theta^2}$$

$$\ell''(\theta) = \frac{n}{2\theta^2} - \frac{\Sigma(y_i-\mu)^2}{\theta^3}.$$

The ML estimate of θ is

$$\hat{\theta} = \frac{1}{n}\Sigma(y_i-\mu)^2$$

and

$$I_{\hat{\theta}} = n/2\hat{\theta}^2.$$

In large samples $(\theta-\hat{\theta})I_{\hat{\theta}}^{\frac{1}{2}}$ has a standard normal distribution, but this is a very poor approximation in small samples because the distribution of $n\hat{\theta} = \Sigma(y_i-\mu)^2$, namely $\theta\chi_n^2$, is heavily skewed.

These properties are again visible in the (relative) likelihood function. We have

$$R(\theta) = L(\theta)/L(\hat{\theta})$$
$$= \left(\frac{\hat{\theta}}{\theta}\right)^{n/2} \exp \left(-\tfrac{1}{2}\left[\frac{\Sigma(y_i-\mu)^2}{\theta} - n\right]\right)$$

A plot of $R(\theta)$ for $n = 10, \Sigma(y_i-\mu)^2 = 10$ shows extreme skew, and it is clear that we cannot treat $(\hat{\theta}-\theta)I_{\hat{\theta}}^{\frac{1}{2}}$ as $N(0,1)$ for the purpose of constructing confidence intervals for θ. Note that $R(\theta)$ is, apart from a term in $\hat{\theta}$, the $\theta\chi_n^2$ density of $\Sigma(y_i-\mu)^2$. The approximating normal likelihood

$$R_A(\theta) = \exp -\tfrac{1}{2}(\theta-\hat{\theta})^2 I_{\hat{\theta}}$$

can also be plotted, and looks very different.

There is nothing to prevent us, however, from reparametrising the model to $\phi = g(\theta)$. Suppose $\sigma = \theta^{\frac{1}{2}}$ is regarded as the parameter instead of σ^2. Then

$$R(\sigma) = \left(\frac{\hat{\sigma}}{\sigma}\right)^n \exp \left(-\tfrac{1}{2}\left[\frac{\Sigma(y_i-\mu)^2}{\sigma^2} - n\right]\right)$$

Plotting the reparametrised relative likelihood for any parameter is very simple in GLIM as we simply calculate $R(\theta)$ and the appropriate function $g(\theta)$ of θ, and plot $R(\theta)$ against $g(\theta)$. The relative likelihood for σ is much less skewed than that for $\theta = \sigma^2$, though it is still far from normal. If $\phi = \log\sigma$ is taken as the parameter, the skewness is further reduced. Using $\psi = \sigma^{-1}$ removes it almost completely. Is there a transformation of σ which makes the likelihood normal (almost exactly)? Anscombe (1964) derived a family of transformations which remove the cubic term in the log-likelihood, so the main component of skewness is eliminated. For the normal variance, the transformation is $\delta = \sigma^{-2/3}$, very close to σ^{-1}. The likelihood function in δ is indistinguishable from normal in samples of n = 10.

This result has striking consequences. Suppose we want a 95% confidence interval for σ^2, given n = 10 and $\Sigma(y_i-\mu)^2 = 10$. The interval based on equal upper- and lower-tail areas of the χ^2_{10} distribution is well known to be biased; it is (0.49, 3.08). The shortest unbiased confidence interval is (0.46, 2.84). If we use the normal approximation to the likelihood function in δ, the approximate 95% confidence interval for δ is $\hat{\delta} \pm 1.96 \ (9\hat{\delta}^{4/3}/2)^{-\frac{1}{2}}$ or (1 ± 0.292), or $(.708, 1.292)$. The corresponding interval for $\sigma^2 = \delta^{-3}$ is $(0.464, 2.818)$, which agrees very closely with the shortest unbiased confidence interval found above. But now we note that we could have obtained this interval without using the normalising transformation at all, because the 95% confidence interval for δ is a 14.6% likelihood interval, and the same interval for σ^2 could have been obtained directly as the 14.6% likelihood interval from the relative likelihood for $\theta = \sigma^2$, or for that matter, from the likelihood for any other monotone function of σ^2.

We complete this section with a comment about the shortest unbiased confidence interval. Why use this rather than the conventional equal-tailed interval? The reason is that its bias, and non-shortest property, arise directly from the fact that the equal-tailed interval is not a likelihood interval, and the shortest unbiased confidence interval is in fact a likelihood interval (Kendall and Stuart 1973, p.212).

Thus the existence of a normalizing transformation for a parameter has very important implications. The 14.6% likelihood interval for σ^2 has a confidence coefficient in repeated sampling of almost exactly 95%, even though the likelihood function in σ^2 is very far from normal. Because a normalizing transformation $g(\sigma^2)$ exists, we know that the likelihood for $\delta = g(\sigma^2)$ is almost exactly normal, and that therefore on this transformed scale the usual "large-sample" confidence interval based on the normal approximation to the likelihood will have the correct confidence coefficient, even in quite small samples. The transformation of this interval back to the original scale - that is, the likelihood interval on the original scale - will then have the same confidence coefficient.

How small does the sample size have to be before the normal approximation breaks down? A value of n = 2 would not usually be thought large, but the approximation works very well even in this case.

2.3 Both μ and σ unknown

In practice μ and σ are both unknown, and we usually want to draw an inference about μ with σ a nuisance parameter. First we consider the case where both μ and σ are of interest.

The likelihood function in both parameters is maximized at $\hat{\mu} = \bar{y}$, $\hat{\sigma}^2 = \frac{1}{n}\Sigma(y_i-\bar{y})^2$, and the relative likelihood function $R(\mu,\sigma)$ is

$$R(\mu,\sigma) = L(\mu,\sigma)/L(\hat{\mu},\hat{\sigma}) = \exp(-\frac{n(\bar{y}-\mu)^2}{2\sigma^2}) \cdot \left(\frac{\hat{\sigma}}{\sigma}\right)^n \exp\left(-\frac{n}{2}\left(\frac{\hat{\sigma}^2}{\sigma^2}-1\right)\right)$$

What confidence interpretation, in repeated sampling, can be attached to the likelihood region in μ and σ defined by $R(\mu,\sigma) \geq \gamma$? The first term in $R(\mu,\sigma)$ is a normal likelihood, and the second term can be made almost exactly normal by the reparametrisation $\delta = \sigma^{-2/3}$. Then $-2 \log R(\mu,\sigma)$ is the sum of squares of two independent variables, one exactly normal, the other almost normal, and hence has very nearly a χ_2^2 distribution in repeated sampling. Thus the confidence coefficient of the region $R(\mu,\sigma) \geq \gamma$, i.e. $-2 \log R(\mu,\sigma) \leq -2 \log \gamma$, is $1-\alpha = P(\chi_2^2 \leq -2 \log \gamma)$, or $\gamma = \exp(-\frac{1}{2} \chi_{2,\alpha}^2)$. Thus if $\gamma = .05$, a 5% likelihood region has confidence coefficient given by $1-\alpha = P(\chi_2^2 \leq 5.99) = .95$.

Such regions look awkward if we want to make separate interval statements about μ and σ separately. The theory of simultaneous intervals (Miller 1966) gives simple methods for constructing such intervals, using projections of the region onto the μ and σ axes. Any function of μ and σ (e.g. $\mu + k\sigma$ or $\mu + k\sigma^2$) can be handled in the same way.

This interval construction can be formalised by considering the profile likelihood for each parameter. If $L(\theta,\phi)$ is a two-parameter likelihood, the profile likelihood of θ is defined as

$$PL(\theta) = L(\theta,\hat{\phi}(\theta))$$

where $\hat{\phi}(\theta)$ is the solution of $\partial L/\partial \phi = 0$. Then the profile likelihood interval for θ based on $PL(\theta) \geq \gamma$ is exactly the projection onto the θ-axis of the likelihood region $R(\theta,\phi) \geq \gamma$. The confidence coefficient of the profile likelihood interval in repeated sampling is then at least $1-\alpha$, where as above $\gamma = \exp(-\frac{1}{2} \chi_{2,\alpha}^2)$.

This looks complicated, but is not. Let us construct the profile likelihood for μ. We have

$$\ell(\mu,\sigma) = -n\log\sigma - \frac{1}{2\sigma^2}\left[\Sigma(y_i-\bar{y})^2 + n(\bar{y}-\mu)^2\right]$$

$$\frac{\partial\ell}{\partial\sigma} = -\frac{n}{\sigma} + \frac{1}{\sigma^3}\left[\Sigma(y_i-\bar{y})^2 + n(\bar{y}-\mu)^2\right]$$

The solution of $\partial\ell/\partial\sigma = 0$ is

$$\hat{\sigma}^2(\mu) = \frac{1}{n}\left[\Sigma(y_i-\bar{y})^2 + n(\bar{y}-\mu)^2\right]$$

Then

$$PL(\mu) = L(\mu,\hat{\sigma}(\mu)) = \frac{1}{\hat{\sigma}^n(\mu)} e^{-n/2}$$

and the profile relative likelihood is

$$PR(\mu) = L(\mu,\hat{\sigma}(\mu))/L(\hat{\mu},\hat{\sigma})$$

$$= \frac{\hat{\sigma}^n}{\hat{\sigma}^n(\mu)}$$

$$= \frac{1}{\left(1 + \frac{n(\bar{y}-\mu)^2}{\Sigma(y_i-\bar{y})^2}\right)^{n/2}}$$

$$= \frac{1}{\left(1 + \frac{t^2}{n-1}\right)^{n/2}}$$

with

$$t^2 = \frac{n(\bar{y}-\mu)^2}{\Sigma(y_i-\bar{y})^2/(n-1)} = \frac{n(\bar{y}-\mu)^2}{s^2}$$

Apart from a multiplicative constant, the relative likelihood has the form of a t_{n-1} density, and a $100\gamma\%$ profile likelihood interval $PR(\mu) \geq \gamma$ is equivalent to

$$\mu \in \bar{y} \pm c\,s/\sqrt{n}$$

with

$$c^2 = (n-1)(\gamma^{-2/n} - 1).$$

The corresponding profile likelihood for σ is obtained from

$$\frac{\partial \ell}{\partial \mu} = \frac{n}{\sigma^2}(\bar{y}-\mu) = 0$$

giving

$$\hat{\mu}(\sigma) = \bar{y} = \hat{\mu},$$

so that

$$PL(\sigma) = L(\hat{\mu}(\sigma),\sigma)$$

$$= \frac{1}{\sigma^n} \exp\left(-\tfrac{1}{2}\frac{\Sigma(y_i-\bar{y})^2}{\sigma^2}\right)$$

and

$$PR(\sigma) = \left(\frac{\hat{\sigma}}{\sigma}\right)^n \exp\left(-\tfrac{1}{2}\left[\frac{\Sigma(y_i-\bar{y})^2}{\sigma^2} - 1\right]\right)$$

2.4 σ a nuisance parameter

In practice we want to make an inferential statement about μ only, treating σ as a nuisance parameter. In the general case of a vector parameter $\underline{\theta}$ with $\underline{\theta}' = (\underline{\theta}_1',\ \underline{\theta}_2')$, suppose that $\underline{\theta}_1$ are the parameters of interest and $\underline{\theta}_2$ are nuisance parameters. In large sample maximum likelihood theory, statements about $\underline{\theta}_1$ are based on the asymptotic $\chi^2_{k_1}$ distribution of $(\hat{\underline{\theta}}_1 - \underline{\theta}_1)'\hat{\mathcal{I}}^{11}(\hat{\underline{\theta}}_1 - \underline{\theta}_1)$ where

$$\hat{\mathcal{I}}^{11} = I_{11} - I_{12}\,I_{22}^{-1}\,I_{21}$$

$$I_{\underline{\theta}} = \begin{bmatrix} I_{11} & I_{12} \\ I_{21} & I_{22} \end{bmatrix}$$

It is easily verified that if the likelihood in $\underline{\theta}$ is multivariate normal:

$$R(\theta) = \exp\{-\tfrac{1}{2}(\underline{\theta} - \hat{\underline{\theta}})\Sigma^{-1}(\underline{\theta} - \hat{\underline{\theta}})\}$$

then the profile relative likelihood in $\underline{\theta}_1$ is

$$PR(\underline{\theta}_1) = \exp\{-\tfrac{1}{2}(\underline{\theta} - \hat{\underline{\theta}}_1)\Sigma^{11}(\underline{\theta} - \hat{\underline{\theta}}_1)\}.$$

Thus the confidence coefficient attached to the profile likelihood region $PR(\theta_1) \geq \gamma$ is given by $\gamma = \exp(-\tfrac{1}{2}\chi^2_{k_1,\alpha})$ in large samples.

However, this confidence coefficient may not be accurate even when the likelihood function in $\underline{\theta}$ is nearly normal. In the normal mean example above, we have

$$I_{\hat{\mu},\hat{\sigma}} = \begin{bmatrix} n/_{\hat{\sigma}^2} & 0 \\ 0 & 2n/_{\hat{\sigma}^2} \end{bmatrix}$$

so that the large-sample result is $(\frac{n(\bar{y}-\mu)}{\hat{\sigma}})^2 \sim \chi^2_1$. But this is not adequate except for sample sizes of $n > 30$, even though the joint likelihood in (μ,σ) is normalisable down to $n = 2$. The reason is that the likelihood in (μ,δ), though the product of independent normal likelihoods, is not bivariate normal, because δ is a scale parameter in one component and a location parameter in the other. The exact profile likelihood in μ is a t_{n-1} likelihood, as we saw above.

3. MODEL CHOICE IN UNBALANCED CROSS CLASSIFICATIONS

How does all this relate to the choice of a regression model for cross-classifications? The choice of an appropriate parsimonious model for cross-classifications remains a contentious issue, and the simultaneous inference approach I proposed in Aitkin (1978, 1979, 1980) has been considered, but not generally adopted, by many writers, particularly in the field of contingency table analysis. To simplify matters I will leave aside the additional difficulties of contingency tables; they will be considered elsewhere, Aitkin (1982). The Quine data used in Aitkin (1978) will provide an illustration.

Given the normal regression model $\underline{y} \sim N(X\underline{\beta}, \sigma^2 I)$, our aim is to make inferential statements about $\underline{\beta}$, with σ a nuisance parameter. The profile likelihood function in $\underline{\beta}$ is easily obtained:

$$L(\underline{\beta}, \sigma) = \frac{1}{\sigma^n} \exp\left(-\frac{1}{2\sigma^2} (\underline{y} - X\underline{\beta})'(\underline{y} - X\underline{\beta})\right)$$

$$\hat{\sigma}^2(\underline{\beta}) = \frac{1}{n}(\underline{y} - X\underline{\beta})'(\underline{y} - X\underline{\beta}) = \frac{1}{n} RSS(\underline{\beta})$$

where $RSS(\underline{\beta})$ is the residual sum of squares from the model for the given value of $\underline{\beta}$. Then

$$PL(\underline{\beta}) = \frac{n^{n/2}}{(RSS(\underline{\beta}))^{n/2}} e^{-n/2}$$

and the profile relative likelihood is

$$PR(\underline{\beta}) = \left\{\frac{RSS}{RSS(\underline{\beta})}\right\}^{n/2}$$

where

$$\hat{\underline{\beta}} = (X'X)^{-1}X'\underline{y}$$

is the maximum likelihood estimate of $\underline{\beta}$, and $RSS = RSS(\hat{\underline{\beta}})$. Straightforward manipulation leads to

$$PR(\underline{\beta}) = \left\{1 + \frac{(\hat{\underline{\beta}} - \underline{\beta})'X'X(\hat{\underline{\beta}} - \underline{\beta})}{RSS}\right\}^{-n/2}$$

so that a profile likelihood region in $\underline{\beta}$ given by $PR(\underline{\beta}) \geq \gamma$ is equivalent to

$$(\hat{\underline{\beta}} - \underline{\beta})'X'X(\hat{\underline{\beta}} - \underline{\beta}) \leq RSS(\gamma^{-2/n} - 1)$$

Since in repeated sampling the distribution of

$$\frac{(\hat{\underline{\beta}} - \underline{\beta})'X'X(\hat{\underline{\beta}} - \underline{\beta})/k}{RSS/(n-k)}$$

is $F_{k, n-k}$, the confidence coefficient $1-\alpha$ attached to the profile likelihood region is given by

$$\gamma^{-2/n} = 1 + kF_{k, n-k}^{\alpha}/(n-k).$$

For the Quine data, with $n = 146$, $k = 28$, this relation is shown below, for several values of α.

α	.75	.50	.25	.10	.05	.025	.01	.005
conf.coeff 1-α	.25	.50	.75	.90	.95	.275	.99	.995
γ	3.1×10^{-6}	2.3×10^{-7}	1.1×10^{-8}	6.2×10^{-10}	9.1×10^{-11}	1.6×10^{-11}	1.8×10^{-12}	4.3×10^{-13}
log γ	-12.7	-15.3	-18.3	-21.2	-23.1	-24.9	-27.0	-28.5

α	.001
conf.coeff 1-α	.999
γ	1.6×10^{-14}
log γ	-31.8

Now consider any arbitrary model for the data, whether determined by a hierarchical reduction of the saturated model, by non-hierarchical setting of main effect
parameters to zero while retaining interactions, by eye inspection of the parameter
estimates, or by any other method. Let $\tilde{\beta}$ be the value of β for this model (some
components of $\tilde{\beta}$ may be set to specific values like zero, and other components may be
estimated), and $RSS(\tilde{\beta})$ the residual sum of squares for this model. The profile relative likelihood for this model is

$$PR(\tilde{\beta}) = \{\frac{RSS}{RSS(\tilde{\beta})}\}^{n/2}$$

This is very easily calculated, and the profile likelihood value for the model plotted
on a log scale against the number of parameters, giving a diagnostic plot similar to
Mallows' C_p-plot or Spjøtvoll's (1977) alternative P_p plot. The vertical scale can
also carry the confidence coefficient attached to the profile likelihood.

For the Quine data, we consider a number of models discussed in Aitkin (1978),
using the original scale. The models, residual sums of squares, number of non-zero
parameters and log profile relative likelihood, are shown below

MODEL	RSS	number of parameters	log PR
C.S.A.L. (saturated)	23,363	28	0.00
no 4-way interaction	23,559	26	-0.61
no 3-way or 4-way interactions	25,939	18	-7.63
all main effects	32,099	7	-23.2
(C + S).A	28,038	12	-13.3
L + (C+S).A	26,933	13	-10.4
(A + S).C	29,784	10	-17.7
(A + L).C	29,448	9	-16.9
(C + L).A	29,264	11	-16.4
A.S. + C.L.	29,189	11	-16.3
C + (S + L).A	28,977	12	-15.7
L + (A + S).C	29,083	12	-16.0

All the models tested, apart from the main effects model, are contained in the 75% confidence region for $\underline{\beta}$.

REFERENCES

Aitkin, M. (1978). The analysis of unbalanced cross-classifications (with Discussion) J.Roy. Statist.Soc.A 141 195-223.

Aitkin, M. (1979). A simultaneous test procedure for contingency table models. Appl.Statist 28 233-242.

Aitkin, M. (1980). A note on the selection of log-linear models. Biometrics 36 173-178.

Aitkin, M. (1982). Logit models for the contigency table analysis of large-scale survey data: the direct likelihood interpretation of G^2 for goodness-of-fit in large sparse tables, and its implication for model selection. Interface XIV Proceedings (to appear).

Anscombe, F.J. (1964). Normal likelihood functions. Ann.Inst.Statist.Math. 26 1 - 19.

Barnard, G.A.(1949). Statistical inference (with Discussion). J.Roy.Statist.Soc.B 13 46-64.

Barnard, G.A., Jenkins, G.M. and Winsten, C.B.(1962). Likelihood inference and time series (with Discussion). J.Roy.Statist.Soc.A 125 321-372.

Edwards, A.W.F. (1972). Likelihood. Cambridge University Press.

Fisher, R.A. (1912). On an absolute criterion for fitting frequency curves. Mess. Math. 41 155-160.

Fisher, R.A. (1934). Two new properties of mathematical likelihood. Proc.Roy.Soc.A 144 285-307.

Kalbfleisch, J.D. and Sprott, D.A. (1970). Application of likelihood methods to models involving large numbers of parameters (with Discussion). J.Roy.Statist Soc.B 32 175-208.

Kendall, M.G. and Stuart, A. (1973). The Advanced Theory of Statistics. Vol.3. (3rd edn.) New York : Hafner.

Miller, R.G. (1966). Simultaneous Statistical Inference. New York : McGraw-Hill

Spjøtvoll, E. (1977). Alternatives to plotting C_p in multiple regression. Biometrika 64 1 - 8.

Sprott, D.A. and Kalbfleisch, J.D. (1969). Examples of likelihoods and comparisons with point estimates and large sample approximations. J.Amer.Statist.Assoc 64 468-484.

Score Tests in GLIM with Applications

Daryl Pregibon

Bell Laboratories
Murray Hill, New Jersey 07974

Keywords:
Efficient scores; Likelihood ratios; Residual analysis; Goodness-of-fit;
Variable selection; Mantel-Haenszel tests.

ABSTRACT

The most common method of hypothesis testing in GLIM is the likelihood ratio method. However, in certain biostatistical application areas, score tests are more commonly used. Mantel-Haenszel chi-squared tests provide good examples. In other cases where a large number of competing models are being entertained, score tests may also be preferable for economy in computing.

We show that score tests can be computed in GLIM with the same ease as likelihood ratio tests. This allows flexibility to users which was not otherwise available. The method is applied to several examples in order to illustrate its usefulness and generality.

1. Introduction

The Generalized Linear Interactive Modelling (GLIM) system was developed with a heavy reliance on the theory of maximum likelihood. According to this theory, there are three common ways of testing hypotheses concerning the regression coefficients β in a generalized linear model (glm). These methods are based on:

M1 - the asymptotic normal distribution of $\hat{\beta}$;

M2 - the asymptotic chi-squared distribution of the likelihood ratio; and

M3 - the asymptotic normal distribution of the score function.

Depending upon the particular application, any of M1-M3 may be preferred. For hypothesis tests concerning single coefficients, M1 is usually used since further fitting is not required. For hypothesis tests concerning several coefficients, M2 and M3 are usually used. For distributions other than the normal, the former is computationally more intensive than the latter since iteration is required during the fitting process. In the biostatistical literature, score tests based on M3 have historically been used for testing purposes, *e.g.* Mantel-Haenszel tests.

The GLIM system was designed to facilitate the use of the likelihood ratio method for testing hypotheses. A generalization of "variance", called "deviance", was introduced for this purpose. In this paper we demonstrate that with similar facility, the method of efficient scores can be used for testing hypotheses in GLIM. The fact that GLIM routinely offers both estimates and hypothesis tests makes it an attractive environment for analyzing data in many disciplines.

The notation used in the development of the score test is set out in Section 2. The use of GLIM to compute the test statistic is discussed in Section 3. Sections 4-7 illustrate the methodology for residual analysis, goodness-of-fit assessment, variable selection, and significance testing in a matched case-control study of cigarette smoking among asbestos miners.

2. Generalized Linear Models

Consider data of the form $\{y_i, \mathbf{x}_i; i=1,...,N\}$ where y is the response variable of interest and $\mathbf{x}=(x_1, x_2, \ldots, x_m)^T$ is a vector of explanatory variables or covariates. A generalized linear model (Nelder and Wedderburn, 1972) is defined for these data by specifying the random and systematic components of the response:

$f(y;\theta)$ - the error distribution which we assume is a member of the one parameter exponential family
$\quad f(y;\theta) = \exp[y\theta - a(\theta) + b(y)]$; and

$g(\mu)$ - the link function which relates μ, the mean of y, to the explanatory variables via $g(\mu) = \mathbf{x}^T\beta$.

The parameters θ and μ are functionally related through $\mu = \dfrac{d}{d\theta} a(\theta)$. Inverting this relationship gives θ as a particular link function $\theta = g_n(\mu)$, called the natural exponential family link function. To simplify notation we will often use this link function in our presentation. It should be noted however that the methods directly carry over to the more general case.

Generalized linear models are conveniently fitted by the method of maximum likelihood. For natural exponential family link functions, the log likelihood function of β is given as

$$L(\beta;\mathbf{y}) = \sum_{i=1}^{N} y_i \mathbf{x}_i^T\beta - a(\mathbf{x}_i^T\beta) + b(y_i).$$

In an obvious vector notation, the score vector and information matrix are given as

$$\mathbf{u}(\beta) = \mathbf{X}^T(\mathbf{y}-\mu) = \mathbf{X}^T\mathbf{s} \quad \text{and} \quad \mathbf{A}(\beta) = \mathbf{X}^T/\text{var } y/\mathbf{X} = \mathbf{X}^T\mathbf{V}\mathbf{X}.$$

The fitting algorithm in GLIM proceeds by updating values of β using

$$\beta_{t+1} = \beta_t - \mathbf{A}(\beta_t)^{-1}\mathbf{u}(\beta_t) = (\mathbf{X}^T\mathbf{V}_t\mathbf{X})^{-1}\mathbf{X}^T\mathbf{V}_t\mathbf{y}_{\tau_t}$$

where $\mathbf{y}_{\tau_t} = \mathbf{X}\beta_t + \mathbf{V}_t^{-1}\mathbf{s}_t$ is the "working vector". When convergence is obtained, we denote the maximum likelihood estimate (mle) as $\hat{\beta}$ and use $\mathbf{V}$ and $\mathbf{s}$ (without hat's) to denote the values of these quantities evaluated at $\hat{\beta}$.

An estimate of the asymptotic covariance matrix of $\hat{\beta}$ is provided by $\mathbf{A}^{-1}$. Hypothesis tests using M1 rely on this matrix. Two measures which summarize the fit are the chisquared and deviance statistics given respectively by

$$X^2 = (\mathbf{y}_\bullet - \mathbf{X}\hat{\beta})^T \mathbf{V}(\mathbf{y}_\bullet - \mathbf{X}\hat{\beta}) \quad \text{and} \quad D^2 = -2[L(\hat{\beta};\mathbf{y}) - L(\hat{\theta};\mathbf{y})],$$

where $\hat{\theta}$ is the N-dimensional mle of θ which fits each point exactly. In addition to their usefulness as summary measures, D^2 and X^2 also form the basis for testing using methods M2 and M3 respectively. Differencing deviances results in likelihood ratio tests. In the next section we show that differencing chisquareds results in score tests.

3. Score Tests in GLIM

Consider a generalized linear model with systematic linear component

$$g(\mu_F) = \mathbf{X}_p\beta_p + \mathbf{X}_q\beta_q. \tag{1a}$$

In this expression, the subscripts p and q refer to subsets of the set of explanatory variables $\mathbf{X}$ of size p and q respectively.

A test of the hypothesis $H:\beta_q=0$ is easily carried out in GLIM using the likelihood ratio method. In addition to a fit of the full model (1a), this method requires a fit of the reduced model:

$$g(\mu_R) = X_p\beta_p. \tag{1b}$$

The likelihood ratio test statistic is thus

$$D^2(\hat{\mu}_R,\hat{\mu}_F) = D^2(\hat{\mu}_R,y) - D^2(\hat{\mu}_F,y)$$

where $D^2(\hat{\mu}_A,y)$ is the deviance from the model with systematic linear component $g(\mu_A)$. Apart from a scale factor present in certain glm's, the asymptotic distribution of $D^2(\hat{\mu}_R,\hat{\mu}_F)$ is approximately $\chi^2(q)$.

Alternatively, one may consider the score test (Rao, Section 6e, 1973) of the hypothesis $H:\beta_q=0$ given by

$$S^2(\hat{\mu}_R,\hat{\mu}_F) = u(\hat{\beta}_R)^T A(\hat{\beta}_R)^{-1}u(\hat{\beta}_R)$$

where $\hat{\beta}_R=(\hat{\beta}_q,0)$. The asymptotic distribution of $S^2(\hat{\mu}_R,\hat{\mu}_F)$ is also approximately $\chi^2(q)$. Note that although S^2 is given an argument μ_F, the statistic (apparently) depends only on a fit of the reduced model (1b).

We now describe how the score test can be computed in GLIM. Evaluation of the first and second log likelihood derivatives at the reduced model mle $\hat{\beta}_R$ leads to

$$S^2(\hat{\mu}_R,\mu_F) = s^T X_q(X_q^T V X_q - X_q^T V X_p(X_p^T V X_p)^{-1}X_p^T V X_q)^{-1}X_q^T s$$

$$= s^T X_q(X_q^T V^{\frac{1}{2}}M_p V^{\frac{1}{2}}X_q)^{-1}X_q^T s$$

where $M_p=I-V^{\frac{1}{2}}X_p(X_p^T V X_p)^{-1}X_p^T V^{\frac{1}{2}}$. Note that both V and s are evaluated at $\hat{\beta}_R$. In addition, let y_* denote the "working vector" from the reduced model. Then, $V^{-\frac{1}{2}}s=M_p V^{-\frac{1}{2}}s$ and $V^{-\frac{1}{2}}s=V^{\frac{1}{2}}(y_*-X\hat{\beta}_R)$, implying that $V^{-\frac{1}{2}}s=M_p V^{\frac{1}{2}}y_*$, leading to an equivalent expression for the score statistic as

$$S^2(\hat{\mu}_R,\mu_F) = y_*^T V^{\frac{1}{2}}M_p V^{\frac{1}{2}}X_q(X_q^T V^{\frac{1}{2}}M_p V^{\frac{1}{2}}X_q)^{-1}X_q^T V^{\frac{1}{2}}M_p V^{\frac{1}{2}}y_*.$$

Readers familiar with hypothesis testing in linear models may recognize S^2 as the difference in residual sum of squares between models

$$y_* = X_p\beta_p + X_q\beta_q + e, \quad e \approx N(0,V^{-1}) \tag{2a}$$

and

$$y_* = X_p\beta_p + e, \quad e \approx N(0,V^{-1}). \tag{2b}$$

That is, the score statistic is the additional sum of squares due to the hypothesis $H:\beta_q=0$ in the weighted linear regression of y_* on X. As the residual sum of squares from the weighted linear regression model (2b) is equal to $X^2(\hat{\mu}_R,y)$ from the generalized linear model (1b), the only additional computation required for the score test is the residual sum of squares (RSS) from the weighted linear regression model (2a). However, as this fit corresponds to the first iteration (starting from $\hat{\beta}_R$) toward the maximum likelihood estimate of β_F in the generalized linear model (1a), we have $RSS=X^2(\hat{\mu}_F^1;y)$ where $\hat{\mu}_F^1$ is the one-step approximation to $\hat{\mu}_F$. Thus the score statistic corresponds to differencing X^2 statistics

$$S^2(\hat{\mu}_R,\hat{\mu}_F^1) = X^2(\hat{\mu}_R,y) - X^2(\hat{\mu}_F^1,y),$$

and contrary to currently held notions, implicitly requires a fit *(albeit* one iteration) of the full model (1a).

In terms of a GLIM analysis, exactly the same number of instructions are required for the score and likelihood ratio tests:

Lik. ratio test	Score test
FIT Xp	FIT Xp
CALC %D=%DV	CALC %C=%X2
REC 10	REC 1
FIT +Xq	FIT +Xq
CAL %D-%DV	CAL %C-%X2

Before concluding this section we remark that the score test works in GLIM because of the way X^2 is computed. In particular, X^2 is "an iteration behind" other quantities which are available for display since it is computed in the SWEEP rather than in the CYCLE routine. This means however that $X^2(\hat{\mu}_R,y)$ may not be exact for the initial fit of the reduced model (1b). Although this will often be the case, its effect is negligible. To ensure that $X^2(\hat{\mu}_R,y)$ is based on $\hat{\beta}_R$ and not an iteration behind, one can compute it as %C=%CU(%WT*(%WV-%LP)**2). This adds one additional step in the computation of S^2. In the examples which follow, we used the value of $X^2(\hat{\mu}_R,y)$ provided by GLIM.

4. Residual Analysis

In the author's unpublished 1979 Ph. D. thesis and in recent articles [see Edwards (1979), DeFize (1980), and Gilchrist (1981)], an analysis of residuals for generalized linear models has been proposed. Attention has centered on standardized residuals defined as

$$t_i = \frac{(y_i - \hat{\mu}_i)/\sqrt{v_i}}{\sqrt{1-h_i}}$$

where $h_i=v_i^{1/2}x_i^T(X^TVX)^{-1}x_iv_i^{1/2}$. In GLIM, h is easily computed as %WT*%VL/%SC.

The above definition is different than the generalized definition of residuals discussed by Cox and Snell (1968) in two important ways:

- t_i treats continuous and discrete variables in an analogous fashion; and

- t_i is defined for censored data as well (see derivation below).

Motivation for the above definition is typically provided by considering approximations to the variance of $(y_i-\hat{\mu}_i)/\sqrt{v_i}$ [see Haberman (1974) or Gilchrist (1981)]. In this section, we provide a more direct derivation via score tests.

Consider a test of the hypothesis that the ith observation is an outlier. If this hypothesis is true, the following model specification is appropriate:

$$g(\mu_j) = x_j^T\beta \qquad \text{for } j \neq i$$

$$g(u_i) = \delta.$$

Alternatively, if z_i denotes an indicator vector such that $z_{ij}=0$ unless $j=i$, this model can be written as

$$g(\mu) = X\beta + z_i\gamma$$

where $\gamma=\delta-x_i^T\beta$. In this parameterization, it is now a simple matter to test the hypothesis that observation i is an outlier since this corresponds to a test of $H:\gamma=0$.

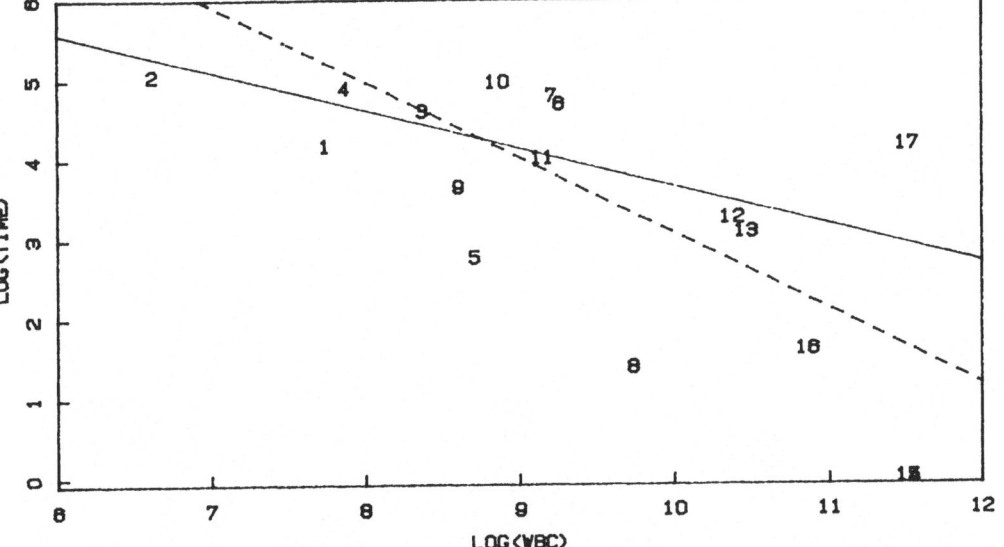

EXHIBIT 1. Scatter plot of the leukemia data: log(survival time) versus log(white blood cell count). Observation number is used as the plotting character. The solid line is the maximum likelihood fit based on the complete data set. The dashed line is the maximum likelihood fit deleting observation seventeen.

Using the notation of sections 2 and 3, the resulting score test statistic can be written as

$$\mathbf{s}^T \mathbf{z}_i (\mathbf{z}_i^T \mathbf{V} \mathbf{z}_i - \mathbf{z}_i^T \mathbf{V} \mathbf{X} (\mathbf{X}^T \mathbf{V} \mathbf{X})^{-1} \mathbf{X}^T \mathbf{V} \mathbf{z}_i)^{-1} \mathbf{z}_i^T \mathbf{s} = \frac{s_i^2 / v_i}{1 - h_i}.$$

Furthermore, since $s_i = y_i - \hat{\mu}_i$, the squared standardized residual t_i^2 is identically the score test statistic corresponding to the hypothesis that observation i is an outlier. The relevant asymptotic distribution of t_i^2 is therefore $\chi^2(1)$. Under an assumption of symmetry, the use of t_i as a standard normal variable is not unreasonable.

As an example we consider some leukemia data analysed by Cox and Snell (1968). The data are plotted in Exhibit 1. The hypothesized model specifies exponential errors and the log link. The solid line corresponds to the maximum likelihood fitted model (ignore dashed line for present). Exhibit 2 displays the $\chi^2(1)$ probability plot of standardized residuals. The null straight line configuration is unreasonable here as observation #17 is out of line. Cox and Snell also identify this observation as having a large residual but conclude that it is consistent with the hypothesized model.

The fact that observation #17 has a large residual and that it also corresponds to a data point with a large x-value, suggests that further analysis is in order. In the author's unpublished Ph. D. thesis, methods for detecting influential data points in generalized linear models were introduced. A summary measure of the coefficient changes due to deleting the ith point is given by

$$c_i^2 = t_i^2 \times \frac{h_i}{1 - h_i}.$$

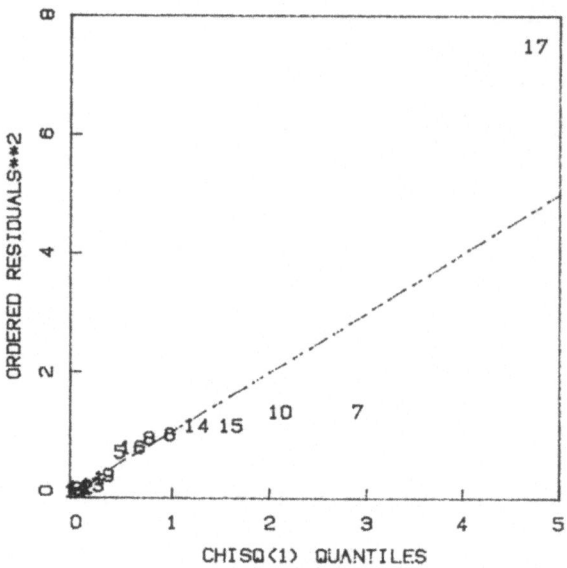

EXHIBIT 2. Chisquared(1) probability plot of squared residuals t_i^2 from the maximum likelihood fit of the leukemia data set. Observation number is used as the plotting character.

This quantity can be derived by considering the effect that deleting the ith observation has on the one-step estimate of $\hat{\beta}$ relative to its asymptotic covariance matrix. An index plot of these quantities is displayed in Exhibit 3. Observation #17 clearly has large influence on the fitted coefficients. The effect on the fitted model is characterized by the dashed line in Exhibit 1. The estimated slope has changed by more than two standard errors. As in linear regression, individual points can have undue influence on a fitted model.

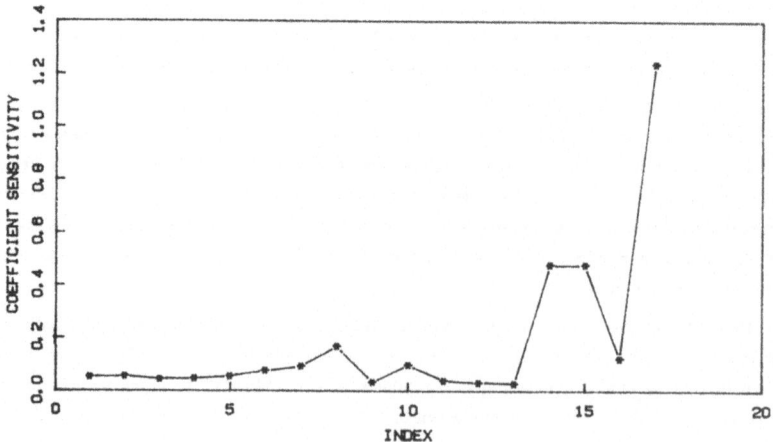

EXHIBIT 3. Index plot of coefficient changes c_i^2 summarizing the effects of individual observation deletion on the maximum likelihood fit of the leukemia data set.

EXHIBIT 4. Listing of GLIM session illustrating the use of the score test in assessing goodness-of-fit.

```
$C READ IN DATA
$UNI 53$DATA XRAY STAGE ACID NODE$DINP 1$
$C SET UP AND FIT THREE VARIABLE LOGISTIC MODEL
$CAL N=1$YVAR NODE$ERROR B N$FIT XRAY+STAGE+ACID$
        scaled
cycle deviance df
   4    50.66    49
$CAL %C=%X2:%D=%DV$
$C CONSTRUCT GROUP FACTOR FOR DECILES OF RISK
$CAL GRP=%GL(11,5):GRP=GRP-%EQ(GRP,11)$FAC GRP 10$
$C REORDER ACCORDING TO SORTED FITTED VALUES
$SORT 1 1 %FV: GRP GRP 1$
$C FIT AUGMENTED MODEL
$REC 1$FIT +GRP$
        scaled
cycle deviance df
   1    34.807   40
--- no convergence by cycle 1
$C CALCULATE SCORE TEST
$CAL %C-%X2$
 15.70
$C ITERATE TO CONVERGENCE
$REC 10$FIT .$
        scaled
cycle deviance df
   7    31.042   40
$C CALCULATE LIKELIHOOD RATIO
$CAL %D-%DV$
 19.62
```

5. Assessing Goodness-of-fit

In a recent paper, Tsiatis(1980) described a goodness-of-fit test for logistic regression models. The method, however, can be applied to any generalized linear model.

Consider a partition of the data points into K disjoint subsets such that points are clustered according to their "nearness" in the covariate space (X). Define the $N \times K$ matrix Z such that $z_{ik}=1$ if observation i is in subset k and 0 otherwise. The adequacy of the hypothesized model

$$g(\mu) = X\beta$$

can be assessed by testing the hypothesis $H:\gamma=0$ in the augmented model

$$g(\mu) = X\beta + Z\gamma.$$

In this formulation it is clear that goodness-of-fit is being assessed by determining whether a separate intercept is required for each "local" subset of data points. The parameter β is maintained to model the "global" effects of the covariates. We have assumed throughout that X contains a constant column, so as stated, the augmented model is over-determined.

The hypothesis $H:\gamma=0$ can be assessed by either a likelihood ratio or score test. Both have an approximate $\chi^2(K-1)$ distribution. Tsiatis suggests using a score test since the resulting test statistic is a quadratic form in "observed minus expected",

$$X^2 = (o-e)^T C^-(o-e).$$

In this equation, o_k and e_k are respectfully the observed and expected totals in the kth subset, and C is the $K \times K$ (singular) covariance matrix of $o-e$. Tsiatis suggests that a generalized inverse of C can be used to compute X^2.

To illustrate the use of GLIM for this application, consider the logistic regression data of Brown(1980) concerning nodal involvement in prostate cancer patients. There are a total of 53 observations and 5 covariates. The fitted model has deviance of 48.13 on 47 df. Two of the covariates do very little in explaining the variation in the response and are omitted in the analysis which follows.

In order to use the method we must first partition the data into meaningful subsets. A standard epidemiological method for this purpose is to group observations according to deciles of estimated risk. Although this is far from optimal (*i.e.* we are grouping according to the model we are attempting to validate) we proceed nonetheless. Exhibit 4 summarizes the GLIM analysis. The score test statistic is 15.70 on 9 df which is nearly significant at the .05 level. In contrast, the likelihood ratio statistic is 19.62 which is significant at the .025 level. Both statistics indicate that further analyses of these data are warranted.

6. Variable Selection

An important component of the generalized linear model specification is the linear predictor $g(\mu) = x^T \beta$. In this section, we are especially interested in the particular variables $x_1, x_2, \ldots, x_m$ which are used to define the linear predictor.

In some applications, it is not uncommon to collect more information per observation than the total number of observations. In these and other less extreme cases, the problem of variable selection is real and must be addressed. Computing costs associated with such a procedure are not trivial, especially when iteration is required to fit each candidate model. Methods which tend to reduce these costs are therefore of some interest. Stepwise selection of variables is one such method.

In a recent paper, Slater (1981) presents a collection of GLIM macros for performing a (forward) stepwise selection of variables for generalized linear models. The criterion he uses is based on the likelihood ratio test, and hence, requires iteration for each candidate model examined. In another recent paper, Peduzzi *et al.* (1980) suggest using score tests in a (forward) stepwise selection procedure for general nonlinear models. Full iteration is performed only when variables enter the model and not for screening candidate models. They argue that computing costs are reduced with negligible effect on associated inferences, that is, the order in which variables enter and when to stop.

The GLIM macros provided by Slater are easily modified to handle score tests. This requires using the normal theory linear model setup of equations (2a & 2b) since successive one-step updates and downdates change the relevant quantities.

EXHIBIT 5. CPU times (seconds) for stepwise fitting using score and likelihood ratio statistics.

n	score	lik. ratio
25	10.1	12.6
50	11.8	17.4
100	15.5	26.9
200	22.9	46.1
400	37.8	83.1

A simple comparison was made to substantiate the advantages of score tests over likelihood ratio tests in selecting variables. For $p=5$ variables and $n=25,50,100,200$ and 400, we performed complete forward stepwise selection using simulated data. The model we used to fit the data assumed gamma errors and a log link. Execution times on a VAX 11/780 under the UNIX* operating system are recorded in Exhibit 5. The tabled values demonstrate the usefulness of score tests even in this limited problem.

EXHIBIT 6. Chi-squared(1) statistics for variable to enter model using score and likelihood ratio statistics.

step	score test variable	χ^2	lik. ratio test variable	χ^2
1	x_2	4.099	x_2	4.154
2	x_5	2.792	x_5	2.825
3	x_3	2.666	x_3	2.691
4	x_1	1.916	x_1	1.933
5	x_4	1.233	x_4	1.243

In order to demonstrate that the gain in computational efficiency is not at the expense of better inferences, we provide values of the χ^2-statistics associated with each step of the analysis for the particular case of $n=200$ (see Exhibit 6). Note in particular that variables enter the model in the same order for both testing methods. In all examples examined to date, this has been the case. Additionally, the test statistics themselves are very similar and are certainly telling the same story concerning the usefulness of adding additional variables to the model. For large data problems, the score test is the way to go.

7. Mantel-Haenszel Tests

Consider a series of 2×2 tables defined by a cross-classification of disease status and exposure:

	Exposed	Not Exposed	Total
With Disease	A_i	B_i	N_{1i}
Free of Disease	C_i	D_i	N_{2i}
Total	M_{1i}	M_{2i}	T_i

Data of this sort arise quite frequently in retrospective studies of disease. In such studies, each table usually represents individuals with common values of variables which the experimenter has decided to control for. A test of the hypothesis that disease is unrelated to exposure is of interest.

Mantel and Haenszel (1959) have proposed the test statistic given by

$$\text{MH} = \frac{\left[\sum (A_i - N_{1i} M_{1i}/T_i) \right]^2}{\sum N_{1i} N_{2i} M_{1i} M_{2i}/T_i^2(T_i-1)}.$$

This test has been widely adopted by practicing biostatisticians and epidemiologists. The optimal properties of the Mantel-Haenszel test are easily derived due to its equivalence with the score test (Day and Byar, 1979). This in turn implies that MH can be computed in GLIM. However, in addition to testing, GLIM conveniently provides an estimate of the log odds ratio and a means of setting confidence intervals. Other aspects of the fit can be routinely examined *(e.g.* a residual analysis to detect outlying tables).

---

* UNIX is a Bell Labs Trademark

The data in Exhibit 7 are extracted from Table 4 of Berry (1980). These data were obtained in a matched (primarily on year of birth) case-control study of the effects of various exposure variables on the mortality of asbestos miners due to lung cancer. In the present analysis, the association between smoking and lung cancer is examined.

EXHIBIT 7. Smoking habits of cases and matched controls in the lung cancer study among asbestos miners.

	Nonsmoking Cases				Smoking Cases		
set	case	control(s)	m	set	case	control(s)	m
1	N	N	1	8	S	N	16
2	N	S	2	9	S	S	19
3	N	NN	4	10	S	NN	14
4	N	NS	5	11	S	NS	24
5	N	SS	4	12	S	SS	21
6	N	NNN	2	13	S	NNN	4
7	N	NSS	2	14	S	NNS	5
				15	S	NSS	7
				16	S	SSS	10
			20				120

There are 16 distinct combinations of smoking status among the 140 matched sets where complete information is available. For example, reference to Exhibit 7 indicates that the seventh set represents the combination of a nonsmoking case and 3 matched controls, 2 of whom are smokers. The column labeled m gives the multiplicity of each distinct combination.

The generalized linear model for these data specifies a logit model for each combination:

$$logit\,(p_j) = \beta_j + z\gamma,$$

where p_j is the probability of dying from lung cancer for an individual in the jth stratum with smoking status z. The hypothesis of no association is $H:\gamma=0$. Due to the matched nature in which the data were collected, a conditional rather than unconditional likelihood analysis is required. GLIM macros for this purpose are presented in the output displayed in Exhibit 8. The score (MH) statistic of 36.74 on one degree of freedom is highly significant. Exhibit 8 also presents results from the fully iterated analysis which are useful for other inferential purposes. In particular, the estimated odds of dying from lung cancer for smokers relative to nonsmokers is $exp(1.71)=5.53$. A 95% confidence interval for the odds ratio is (3.02, 5.53).

EXHIBIT 8. Listing of GLIM session illustrating the score test in case-control studies.

```
$C READ IN DATA
$UNI 50$DAT Y SMOKE M J$DINP 1$
$WRITE MACROS FOR USER DEFINED MODEL
$MAC M1 $CAL E=0:E(J)=E(J)+%EXP(%LP):%FV=%EXP(%LP)/E(J)
     $CAL E=0:E(J)=E(J)+%FV*SMOKE:X=SMOKE-E(J)$END
$MAC M2 $CAL %DR=1/%FV$END
$MAC M3 $CAL %VA=%FV$END
$MAC M4 $CAL %DI=-2*%YV*%LOG(%FV)$END
```

```
$C SETUP AND FIT THE NULL MODEL
$VAR 16 E$OWN M1 M2 M3 M4$SCA 1$WEI M$CAL %LP=0$YVAR Y$FIT $
        scaled
cycle deviance df
   2   294.1   49
$CAL %C=%X2:%D=%DV$
$C FIT AUGMENTED MODEL
$REC 1$CAL X=SMOKE$FIT +X$
        scaled
cycle deviance df
   1   254.5   48
--- no convergence by cycle 1
$C CALCULATE SCORE TEST
$CAL %C-%X2$
36.74
$C ITERATE TO CONVERGENCE
$REC 10$FIT .$
        scaled
cycle deviance df
   2   253.5   48
$C CALCULATE LIKELIHOOD RATIO TEST
$CAL %D-%DV$
40.59
$C EXTRACT PARAMETER ESTIMATES AND VARIANCES
$EXT %PE %VC$CAL %PE(2):%SQRT(%VC(3))$
1.710
0.2309
```

REFERENCES

Berry, G. (1980). Dose-response in case-control studies, Jour. Epid. and Comm. Health **34**, 217-222.

Brown, B.W. (1980). Prediction analyses for binary data, *"Biostatistics Casebook,"* John Wiley and Sons: New York.

Cox, D.R. and Snell, E.J. (1968). A general definition of residuals, Jour. Royal Statist. Soc. B **30**, 248-275.

Day, N.E. and Byar, D.P. (1979). Testing hypotheses in case-control studies - Equivalence of Mantel-Haenszel statistics and logit score tests, Biometrics **35**, 623-630.

DeFize, P.R. (1980). The calculation of adjusted residuals for log-linear models in GLIM, The GLIM Newsletter **3**, 41.

Edwards, D. (1979). Analysis of residuals in two-way contingency tables, The GLIM Newsletter **1**, 30-31.

Gilchrist, R. (1981). Calculation of residuals for all GLIM models, The GLIM Newsletter **4**, 26-28.

Haberman, S.J. (1974). *"The Analysis of Frequency Data,"* University of Chicago Press: Chicago.

Mantel, N. and Haenszel, W. (1959). Statistical aspects of the analysis of data from retrospective studies of disease, Jour. Nat. Cancer Inst. **22**, 719-768.

Nelder, J.A. and Wedderburn, R.W.M. (1972). Generalized linear models, Jour. Royal Statist. Soc. A **135**, 370-384.

Peduzzi, P.N., Hardy, R.J. and Holford, T.R. (1980). A stepwise variable selection procedure for nonlinear regression models, Biometrics **36**, 511-516.

Rao, C.R. (1973). *"Linear Statistical Inference, Second Edition,"* John Wiley and Sons: New York.

Slater, M. (1981). A GLIM program for stepwise analysis, The GLIM Newsletter **4**, 20-25.

Tsiatis, A.A. (1980). A note on a goodness-of-fit test for the logistic regression model, Biometrika **67**, 250-251.

GLIM SYNTAX AND SIMULTANEOUS TESTS FOR GRAPHICAL LOG LINEAR MODELS

Joe Whittaker

Department of Mathematics, University of Lancaster

SUMMARY

Within the class of hierarchical log linear models for contingency tables, conditional independence models have a special place and as they are well represented by their independence graphs they are known as graphical models.   Standard GLIM model formulae are extended to give a simple representation of these models and rules are given to translate formulae between the standard and extended language.   The deviances for all conditional independence models can be computed from the deviances of the elementary graphical models.   These computations are employed in a simultaneous test procedure for selecting a parsimonious model from the class of all graphical models. This procedure is illustrated on a contingency table from the sociological literature classified by four response factors.

Keywords: CONDITIONAL INDEPENDENCE;   CONTINGENCY TABLE;   GRAPHICAL MODEL;   LOG LINEAR MODEL;   SIMULTANEOUS TESTING;   GLIM;   ROLE CONFLICT;   ADEQUATE MODELS.

1. INTRODUCTION

Standard statistical practice for contingency table analysis altered radically during the 1970s.   The development of log linear models made it possible to formulate complex models of the dependences between factors cross classifying the table. Hitherto it had been quite difficult to fit even the most rudimentary of such models.

Within this class of log linear models many can be interpreted in terms of the fundamental probablistic concepts of independence and conditional independence. Goodman (1970) developed a calculus to elicit such interpretations from a given model. See also Bishop et al. (1975).

Two important papers, one by Wermuth (1976) and the other by Darroch et al. (1980), completely reorientated the relationship between conditional independence and log linear models.   They showed that conditional independence is a pairwise relationship between two factors so that in a table cross-classified by k response factors there are $\binom{k}{2}$ conditional independences.   Further, they showed that all conditional independence and independence interpretations are derived from these $\binom{k}{2}$ conditional independences and finally that there are exactly $2^{\binom{k}{2}}$ conditional independence models within the class of log linear models.   Though, there are log linear models which are not conditional independence models but which do have conditional independence interpretations.

Wermuth focussed attention on covariance selection models which are the continuous analogue of conditional independence models and this work was extended in Wermuth (1980). The thrust of Darroch, Lauritzen and Speeds' research was the connection between conditional independence models and Markov fields. One of the several contributions made in that paper is the notion of an independence graph, where vertices correspond to factors and edges to conditional dependences. As the graph of a conditional independence model contains all interpretations of the model in terms of independence and conditional independence a useful synonym for a conditional independence model is a graphical model.

This paper formulates graphical models in the GLIM language by extending the factorial notation of Wilkinson and Rogers (1973) to include the intersection of two model formulae. Model intersection was used by Goodman (1973) and Havranek (1981). This can be used to find the "nearest" graphical model to any given hierarchical log linear model. A simultaneous test procedure is developed along the lines of Aitkin (1979) to select a parsimonious conditional independence model for a four way table concerning role conflict. Only the so called elementary graphical models require fitting. Proofs are omitted here but are included in Whittaker (1982).

2. CONDITIONAL INDEPENDENCE IN LOG LINEAR MODELS

Three or four way tables are usually rich enough to provide the necessary concepts and are simple to deal with notationally. Generalisations to k factors are trivial. So suppose that a table S is cross classified by four factors A,B,C and D and that p is a probability on S. A log linear model is succinctly described in the standard GLIM language by

$$\log p \ \varepsilon \ M$$

where M is a model formulae. Model formulae are constructed from two binary operations between factors, dot and plus, giving A.B and A+B.

It is well known that A and B are independent in a two way table when

$$\log p \ \varepsilon \ A+B$$

and that A and B are conditionally independent given the factor C in a three way table when

$$\log p \ \varepsilon \ (A+B).C$$

2.1 Extended GLIM

These concepts generalise naturally. In a four way table, A and B are conditionally independent given the "rest" if

$$\log p \ \varepsilon \ (A+B).C.D$$

By writing S = A.B.C.D for the saturated model and $\bar{A}$ for the compound factor

$$\bar{A} = B.C.D \qquad\qquad \text{(when S = A.B.C.D)}$$

then A and B are conditionally independent given the rest when

$$\text{logp } \varepsilon \ \bar{A} + \bar{B}$$

This applies to a k way table with S = A.B. .. K by defining $\bar{A}$ = B.C. ... K.

Furthermore if M is any hierarchical log linear model and M $\subset \bar{A}+\bar{B}$ then M contains the conditional independence of factors A and B.   For example

$$\text{logp } \varepsilon \ M = A.B + B.C.D \ \subset A.B.C + B.C.D = \bar{D}+\bar{A}$$

and similarly

$$\text{logp } \varepsilon \ M = A.B + B.C.D \ \subset A.B.D + B.C.D = \bar{C}+\bar{A}$$

Thus factors A and C and factors A and D are conditionally independent in M.

To make further progress we define the binary operation × between model formulae by

Definition : logp ε M×N if and only if logp ε M and logp ε N.

In the above example we have

$$\text{logp } \varepsilon \ (\bar{D}+\bar{A}) \times (\bar{C}+\bar{A})$$

so that A and D are conditionally independent given the rest and A and C are conditionally independent given the rest.   This operation corresponds to the intersection of model formula.   Most importantly it demonstrates that conditional independence is a pairwise relationship, that in a table with k factors there are $\binom{k}{2}$ possible conditional independences and that there are $2\binom{k}{2}$ conditional independence models.

2.2 Model Simplification

Put 1 = $\bar{S}$ = $\overline{A.B.C.D}$.   In standard GLIM this term would correspond to the general mean (%GM).   It is possible to verify the following statements for arbitrary hierarchical models M,N and L defined on S = A.B.C.D.

$$
\begin{array}{lll}
M.M = M & M{\times}M = M & M+M = M \\
M.1 = M & M{\times}1 = 1 & M+1 = M \\
M.S = S & M{\times}S = M & M+S = S \\
\end{array}
$$

$$M{\times}N \subset M \subset M+N \subset M.N$$
$$(M+N).L = M.L + N.L$$
$$(M+N){\times}L = M{\times}L + N{\times}L$$
$$(M.N){\times}L \ne M.(N{\times}L) \qquad \text{in general}$$
$$A{\times}B = 1, \quad \bar{A}{\times}\bar{B} = \overline{A.B}$$

These provide enough ammunition to simplify the conditional independence models.

Example 1. $M = (\bar{D}+\bar{A}) \times (\bar{C}+\bar{A})$

$$= \ \bar{D}{\times}\bar{C} + \bar{A}{\times}\bar{C} + \bar{D}{\times}\bar{A} + \bar{A}{\times}\bar{A}$$

by the distributive rule.   As $\bar{A}\times\bar{C} \subset \bar{A}$ and $\bar{A}\times\bar{A} = \bar{A}$ this simplifies to

$$M = \overline{D.C} + \bar{A}$$

which expresses the conditional independence of A and the compound factor D.C. Finally

$$M = A.B + B.C.D.$$

Example 2.
$$\begin{aligned}
M &= A.B + B.C\ \ + C.D + D.A. \\
&= \overline{C.D} + \overline{A.D}\ \ + \overline{A.B} + \overline{B.C} \\
&= \bar{C}\times\bar{D} + \bar{A}\times\bar{D}\ \ + \bar{A}\times\bar{B} + \bar{B}\times\bar{C} \\
&= (\bar{C}+\bar{A})\times\bar{D} + (\bar{C}+\bar{A})\times\bar{B} \\
&= (\bar{C}+\bar{A})\times(\bar{D}+\bar{B})
\end{aligned}$$

Expressing the conditional independence of A and C and of B and D.   (As an aside note that M = (A+C).(B+D) as well).

3. INDEPENDENCE GRAPHS

Darroch et al (1980) demonstrated how neatly a conditional independence model can be represented by its independence graph.   The graph is constructed with vertices corresponding to the factors and edges to the absence of conditional independences in the model, hence the term graphical model.

Examples

(i) $M = \bar{A}+\bar{B}$

           $= (A+B).C.D$

(ii) $M = (\bar{A}+\bar{C})\times(\bar{A}+\bar{D})\times(\bar{B}+\bar{D})$

           $= A.B + B.C + C.D$

(iii) $M = (\bar{A}+\bar{C})\times(\bar{B}+\bar{D})$

           $= A.B + B.C + C.D + D.A$

There are various methods for checking which, if any, conditional independences are included in a given model.   A simple one is given in Darroch et al (1980) and is illustrated here on an example of Haberman (1974).

Consider a six way table and the log linear model M

$$M = A.E.F + B.D.F + C.D.E$$

Construct the (first order interaction) graph of M by connecting any two vertices of the graph whenever the corresponding two factors occur together in one or more of the summands of M:

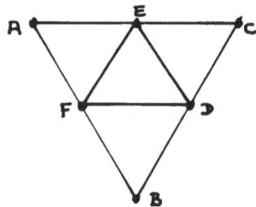

Now write down the conditional independence model, M', as though this was the independence graph, giving

$$M' = (\bar{A}+\overline{B.C.D})\times(\bar{B}+\overline{E.C})\times(\bar{C}+\bar{F})$$

Clearly M⊂M'; the question of equality can only be determined by simplifying M'. Routine computation gives

$$M' = D.E.F + A.E.F + B.D.F + C.D.E = D.E.F + M$$

Hence M is not a conditional independence model, but since M⊂M' it has the same conditional independence interpretations as M.

4. SIMULTANEOUS TESTS

When the factors cross-classifying the table are all response factors and when there is no prior information concerning the pattern of dependence between the factors then a reasonable requirement of a model fitting procedure is to insist it treat the $\binom{k}{2}$ possible conditional independences symmetrically. For even small values of k the number of terms to be considered is quite large. Furthermore, there are several possible tests for each dependence depending on which of the other conditional independences are included in the model. A procedure which controls the overall size of the sequence of tests involved in selecting a parsimonious model   is highly desirable. One such procedure is due to Aitkin (1979); see also Aitkin (1978) and Whittaker and Aitkin (1978).

The saturated model S contains no independences while the minimal model of complete independence I contains all $\binom{k}{2}$. The procedure accepts I if dev(I) < c where c is a predetermined critical value chosen to control the overall STP size. If not choose a particular order to successively include the conditional independences starting with S. Continue until the total deviance attributable to the included independences reaches c but does not exceed it. Then stop; the dependences that still remain in the model are then said to be adequate. This is a minimally adequate model if it does not contain a proper subset of dependences which is also adequate.

The size of this testing procedure is just the probability of rejecting the minimal model I when true. As dev(I) has an asymptotic chisquared distribution with df(I) degrees of freedom then c is the appropriate percentage point from this

table. Now if the test statistics for the $\binom{k}{2}$ conditional independences were independent then

$$P \text{ (type I error)} = 1-(1-\alpha)^{\binom{k}{2}} \simeq \binom{k}{2} \alpha$$

This gives guidance as to the choice of c; in a four way table this is approximately
0.26 when α = .05; hence it might be reasonable to set the overall size at .20, say.

The burden of work in executing this procedure is to determine which subsets of
the $\binom{k}{2}$ conditional independences are minimally adequate. At worst this can be a
fearsome task, for there are $\binom{k}{2}$! possible permutations; some 720 when k = 4 and
roughly 3.6×10^6 when k = 5. The next section reduces this to more manageable
proportions.

This procedure can be extended to a table in which some of the factors are
response factors and some are treatment factors, though a decision to treat the
dependences between response factors and those between response and treatment factors
on an equal footing or not has to be made.

5. ELEMENTARY GRAPHICAL MODELS IN FOUR DIMENSIONS

The graph of $M=\bar{A}+\bar{B}$ is

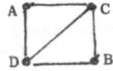

and A and B are separated by C and D. Consider a test for the conditional independence of A and C in this graph; the likelihood ratio test is given by the deviance
difference, d^2, between

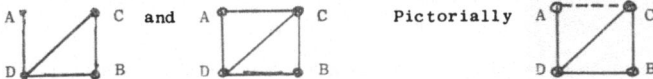

represents the test. Now a consequence of the separation by CD is that the test

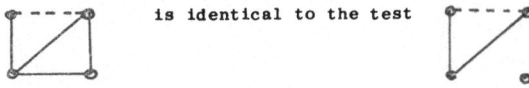

is identical to the test

A proof following by factorising the probability, Whittaker (1982). In fact there
is no need to fit $\bar{A}+\bar{B}$ at all for

$$\text{dev}(\bar{A}+\bar{B}) = \text{dev}(A+\bar{A}) + \text{dev}(B+\bar{B}) - \text{dev}(A+B+\overline{AB})$$

or pictorially

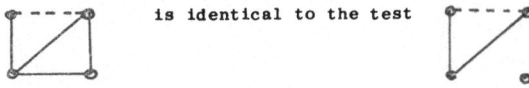

This reduces the maximum number of models that require fitting from 720 to 14. These are the <u>elementary</u> conditional independence models. In four dimensions they are

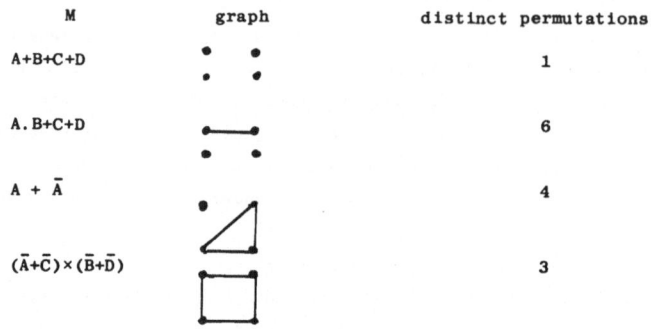

M	graph	distinct permutations
A+B+C+D		1
A.B+C+D		6
A + Ā		4
(Ā+C̄)×(B̄+D̄)		3

6. AN EXAMPLE

To illustrate the simultaneous model fitting of conditional independences we reanalyse a 2^4 contingency table originally published by Stouffer and Toby (1951) and later analysed by Goodman (1974)

TABLE 1.

Role conflict responses of 216 respondents

		C	+		−	
A	B	D	+	−	+	−
+	+		42	23	6	25
	−		6	24	7	38
−	+		1	4	1	6
	−		2	9	2	20

Each respondent was presented with four situations A,B,C and D and the possible responses to each were "+" denoting "universalistic" and "−" denoting "particularistic". Goodman used this example to illustrate techniques of latent structure analysis.

The elementary models were fitted first

TABLE 2.

Deviances for the elementary graphical models

Model	df	dev	Model	df	dev
A+B+C+D	11	81.08	A+Ā	7	20.79
A.B+C+D	10	68.31	B+B̄	7	43.78
A.C+B+D	10	75.33	C+C̄	7	38.24
A.D+B+C	10	71.73	D+D̄	7	49.07
B.C+A+D	10	65.30	(Ā+B̄)×(C̄+D̄)	7	26.94
B.D+A+C	10	56.73	(Ā+C̄)×(B̄+D̄)	7	20.06
C.D+A+B	10	56.35	(Ā+D̄)×(B̄+C̄)	7	15.07

The information in this table can be collated to form a d^2 matrix of deviance differences attributable to pairwise conditional independences. This is similar to the familiar technique of presenting the correlation matrix and the partial correlation matrix together.

TABLE 3

Conditional independence d^2 matrix

Lower triangle, unadjusted, 1 df. Upper triangle, adjusted, 4 df.

	A	B	C	D
A	-	8.22	2.30	4.56
B	12.77	-	10.29	17.52
C	5.75	15.78	-	19.00
D	9.35	24.35	24.73	-

Note that $12.77 = 81.08 - 68.31 = d^2(A.B+C+D, A+B+C+D)$

and $8.22 = 20.79 + 43.78 - 56.35 = d^2(S, \bar{A}+\bar{B})$

The latter figure could have been computed directly by fitting $\bar{A}+\bar{B}$ in GLIM.

6.1 Searching for adequate graphical models.

We set the overall test size to 0.20. As there are 11 df available for testing complete independence, $c = 14.63$. Since $d^2(S, A+B+C+D) = 81.08 > 14.63$ we permute the order of the included independences. With this example a reasonable strategy is to start with the largest adjusted d^2 for conditional independence in Table 3.

$\bar{C}+\bar{D}$: As $19.00 > 14.63$ stop. This independence cannot be included.
The model S is adequate.

$\bar{B}+\bar{D}$: The same.

$\bar{B}+\bar{C}$: Now $10.29 < 14.63$ so this independence can be included.

The adjusted deviance differences are recalculated within

giving

	A	B	C	D
A	-	9.14	3.22	22.20
B		-	*	20.72
C			-	21.80
D				-

Any term with a deviance of less than $14.63 - 10.29 = 4.34$ can be included. There is one, namely $\bar{A}+\bar{C}$. When this is included the adjusted d^2 are recalculated within

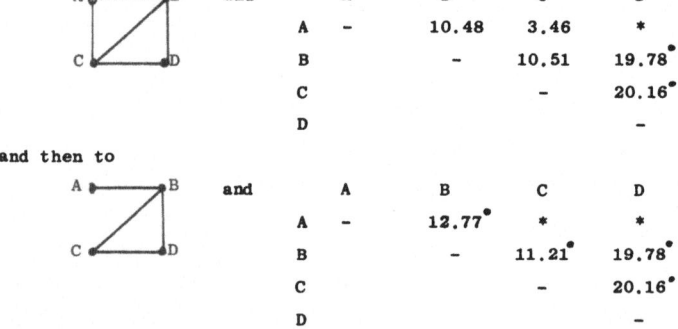

giving

	A	B	C	D
A	–	9.14 •	*	19.23
B		–	*	20.72 •
C			–	24.73 •
D				–

The • denotes that this d^2 has already been calculated, so that only one new calculation was necessary. There are no further independences to include as each d^2 is
greater than 4.34−3.22 = 1.12. Hence M = $(\bar{C}+\bar{A})\times(\bar{C}+\bar{B})$ is an adequate model.

$\bar{A}+\bar{B}$: including this term leads to

with	A	B	C	D
A	–	*	3.62	7.22
B		–	11.21	19.78
C			–	18.72
D				–

The critical value is now 14.63−8.22 = 6.41 so that only $\bar{A}+\bar{C}$ can be included as well.
This gives

with	A	B	C	D
A	–	*	*	9.35 •
B		–	11.21 •	19.78 •
C			–	20.16
D				–

and no further reduction is possible and M = $(\bar{A}+\bar{B})\times(\bar{A}+\bar{C})$ is adequate.

$\bar{A}+\bar{D}$: including this leads to

and	A	B	C	D
A	–	10.48	3.46	*
B		–	10.51	19.78 •
C			–	20.16 •
D				–

and then to

and	A	B	C	D
A	–	12.77 •	*	*
B		–	11.21 •	19.78 •
C			–	20.16 •
D				–

giving M = $(\bar{A}+\bar{D})\times(\bar{A}+\bar{C})$ as adequate.

$\bar{A}+\bar{C}$: the deviance of this model of 2.30 is the smallest. Hence it is the independence most likely to occur in an adequate model and for this reason it has been
considered last. It leads to

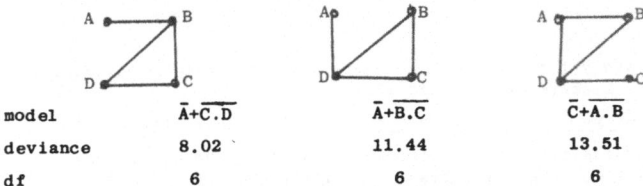

	A	B	C	D
A	–	9.14°	*	5.72
B		–	11.21°	17.76°
C			–	20.16°
D				–

The critical value is now 14.63-2.30 = 12.33 and so either $\bar{A}+\bar{B}$ or $\bar{A}+\bar{D}$ or $\bar{B}+\bar{C}$ can be included next. But each of these models has been tested before and found to be adequate. So that no further computation is necessary.

6.2 Minimally adequate graphical models

The analysis above leads to three minimally adequate models at the 20% significance level:

model	$\bar{A}+\overline{C.D}$	$\bar{A}+\overline{B.C}$	$\bar{C}+\overline{A.B}$
deviance	8.02	11.44	13.51
df	6	6	6

Each contains two conditional independences and four dependences and each contains $\bar{A}+\bar{C}$. All three give a plausible explanation of the dependences in the four way table. Perhaps the first model is to be preferred as its likelihood is larger than that of the others.

Interestingly the deviances for both $\bar{A}+\overline{B.C}$ and $\bar{C}+\overline{A.B}$ are significant at the 20% level when compared to the chisquared table with 6 d.f., though all three models are not significant at the 3.65% level in the same table.

7. CONCLUSION

A comparison of this simultaneous test procedure with a crude analysis of Table 3 is revealing. As the 5% significance level on 1 df is 3.84, all the unadjusted dependences are significant which suggests that only the saturated model is adequate. As the significance level on 4 df is 9.49 inspection of the adjusted deviances lead to the model in which A is conditionally independent of B,C and D and consequently completely independent of them. Neither of these models is acceptable. The correlations between the test statistics underline the necessity for a simultaneous procedure.

Finally, it seems unlikely that the above analysis would have been at all comprehensible without the benefit of the independence graphs and the streamlined notation for log linear models. One might hope that it may soon be possible to automate this search for minimally adequate models.

REFERENCES

AITKIN, M.A. (1978). The analysis of unbalanced cross-classifications (with discussion). J.R. Stat. Soc. A 141, 195-223.

AITKIN, M.A. (1979). A simultaneous test procedure for contingency table models. Appl. Stat. 28, 233-242.

BAKER, R.J. and NELDER, J.A. (1978). The GLIM System. Release 3. N.A.G.

BISHOP, Y., FIENBERG, S. and HOLLAND, P. (1975). Discrete Multivariate Analysis M.I.T. Press.

DARROCH, J., LAURITZEN, S. and SPEED T. (1980). Markov fields and log linear interaction models for contingency tables. Ann. Stat. 8, 522-539.

GOODMAN, L.A. (1970). The multivariate analysis of qualitative data : interaction among multiple classifications. J. Amer. Stat. Soc. 65, 226-256.

GOODMAN, L.A. (1973). The analysis of multidimensional contingency tables when some variables are posterior to others : a modified path analysis approach. Biometrika, 60, 179-192.

GOODMAN, L.A. (1974). Exploratory latent structure analysis using both identifiable and unidentifiable models. Biometrika, 61, 215-232.

HABERMAN, S.J. (1974). The Analysis of Frequency Data. Univ. Chicago Press.

HAVRANEK, T. (1981). On some possibilities of logical analysis of the dependency structure in multidimensional contingency tables. Res. Rep. GN.206. Centre of Biomathematics. Czechoslovak Academy of Sciences, Prague.

STOUFFER, S.A. and TOBY, J. (1951). Role conflict and personality. Amer. J. Sociol. 56, 395-406.

WERMUTH, N. (1976). Model search among multiplicative models. Biometrics 32, 253-263.

WERMUTH, N. (1980). Linear recursive equations, covariance selection and path analysis. J. Amer. Stat. Ass. 75, 963-972.

WHITTAKER, J. (1982). Conditional Independence Models. Unpublished manus.

WHITTAKER, J. and AITKIN, M. (1978). A flexible strategy for fitting complex log linear models. Biometrics, 34, 487-495.

WILKINSON, G.N. and ROGERS, C.E. (1973). Symbolic description of factorial models for analysis of variance. Appl. Stat. 22, 392-399.

COMPOUND POISSON REGRESSION MODELS

John Hinde

Centre for Applied Statistics

University of Lancaster

SUMMARY

Count data are easily modelled in GLIM using the Poisson distribution. However, in modelling such data the counts are often aggregated over one or more factors, or important explanatory variables are unavailable and as a result the fit obtained is often poor. This paper examines a method of allowing for this unexplained variation by introducing an independent random variable into the linear model for the Poisson mean, giving a compound Poisson model for the observed data. By assuming a known form for the distribution of this random variable, in particular the normal distribution, and using a combination of numerical integration, the EM algorithm and iteratively reweighted least squares, maximum likelihood estimates can be obtained for the parameters. Macros for implementing this technique are presented and its use is illustrated with several examples.

KEYWORDS: COUNT DATA; POISSON DISTRIBUTION; OVER-DISPERSION; COMPOUND POISSON; MAXIMUM LIKELIHOOD; NUMERICAL INTEGRATION; EM ALGORITHM; IRLS.

1. INTRODUCTION

Frome et al (1973) described a method of regression analysis for Poisson distributed data, fitting both linear and non-linear regression models in explanatory variables to the Poisson mean using iteratively reweighted least squares (IRLS). At about the same time Nelder and Wedderburn (1972) demonstrated that IRLS could be used to fit a wide class of models, which included a linear model for the log of the Poisson mean, a regression model for count data analogous to the usual normal regression for continuous data. Specifically the Poisson regression model relates an observed response Y to a set of explanatory variables $X_1, \ldots, X_p$ in the following way

$$Y|X_1, \ldots, X_p \sim \text{Pois}(\mu)$$

with $\log(\mu) = \beta_0 + \beta_1 X_1 + \ldots + \beta_p X_p$.

$$(1.1)$$

IRLS provides maximum likelihood estimates for $\beta_0, \beta_1, \ldots, \beta_p$ and this is now most conveniently done in GLIM using

$ERROR POISSON $LINK LOG

and then fitting the appropriate linear model. In the standard GLIM manner the importance of the individual explanatory variables can be assessed using an analysis of deviance for a suitable hierarchy of models, leading to a minimal plausible model involving a subset of the $X_1, X_2, \ldots, X_p$. The goodness of fit of the final model can be assessed by examining the residual deviance, which is the difference in -2log likelihood for the minimal model and that for the complete model which reproduces the data, and which might be expected to be approximately χ^2-distributed on the difference in degrees of freedom in the two models. In addition it is often informative to examine the adjusted residuals, see Haberman (1974) and Gilchrist (1981), where the adjustment refers to a modification of the standardised residual to allow for the variance of the linear predictor. These can be obtained from

$$r_i = \frac{(y_i - \hat{\mu}_i)}{\sqrt{\hat{\mu}_i(1 - \hat{\mu}_i \hat{q}_i)}},$$

where $\hat{\mu}_i$ is the fitted value and $\hat{q}_i$ is the variance of the linear predictor for the ith observation y_i. For an adequate model these can be expected to have, approximately, mean 0 and variance 1, although this result is asymptotic in μ and so not always very helpful.

These Poisson regression methods provide an appropriate means for analysing much count data, which occurs frequently in a wide variety of applications, particularly in geography and the social sciences. Such data typically exhibit a strong mean-variance relationship, and this is often a near-equality relationship, as in the Poisson distribution which is then the appropriate distribution to use as it also reflects the discreteness in the data. If the counts are large, the discreteness ceases to be important, and the Poisson distribution will be well approximated by a normal distribution with variance equal to mean, and in these circumstances the usual weighted least squares methods can be used; this is most easily achieved using weights inversely proportional to the observations rather than the mean, and in practice this is usually adequate, enabling the simple use of a prior weight rather than an iteratively recalculated weight and a user defined model. When some of the counts are small such approximate methods are poor, failing to account for the discreteness of the data, and Poisson methods should be used.

2. COMPOUND POISSON MODELS

2.1 Over-dispersion

A common problem with count data is that, even after allowing for important explanatory variables using the Poisson regression model, the fits obtained are poor. This is reflected in over-large residual deviances and adjusted residuals which have a variance > 1. This indicates that, conditional upon the explanatory variables included in the final model, the variance of an observation is greater than its mean, implying that the Poisson assumption is incorrect. Such data are frequently described as being over-dispersed. There are various ways in which such a phenomenon can arise; for example the response data as presented may have been aggregated, or the usual assumption of independence may be incorrect, i.e. the data are correlated, or simply important explanatory variables may not have been measured and are consequently

incorrectly excluded from the regression relationship. The method to be developed
in this paper will be particularly concerned with the latter problem, although the
method derived will also be of potential use in the other situations, as the actual
mechanism leading to over-dispersed data cannot be determined without additional
information.

2.2 Example

In a study of inter-urban migration on the number of migrants between 126 Standard Metropolitan Labour Areas (SMLAs) data were obtained from the 1971 Census of
Great Britain. The aim was to relate the number of migrants N_{ij}, from region i to
region j, to a number of explanatory variables including the populations of the regions,
P_i, and the distances between regions, d_{ij}. A commonly used model in such studies is
some form of the gravity model

$$E(N_{ij}) = \frac{\alpha P_i^{\beta 1} P_j^{\beta 2}}{d_{ij}^{\gamma}} \ .$$

This fits very easily into the Poisson regression framework, for we just assume the
number of migrants to be Poisson distributed with means given by the gravity model,
which gives a log-linear model for the mean. Fitting such a model to the 15,750
counts of migrants, which range from 0 in over half of the cases to a high of 681,
we obtain a poor fit, with residual deviance approximately 5 times as large as the
residual degrees of freedom. Closer inspection of the problem reveals that rather
than assuming that individuals move independently, according to a gravity model, it
might be more reasonable to expect households to be the important unit of migration.
The total number of people moving from region i to region j can then be expressed as

$$N_{ij} = X_1 + X_2 + \ldots + X_{H_{ij}}$$

where H_{ij}, the number of householders moving from i to j is assumed to have a Poisson
regression model with the mean given by the gravity model, and the X_ℓ are independent
and identically distributed according to the household size distribution in Great
Britain. This distribution is available from the census data and applying such a
model gives a much better fit, indicating that the gravity model may well be appropriate for the number of households moving.

2.3 Compound Poisson Distributions

Compound Poisson distributions can be obtained in a number of ways, for instance
as a Poisson sum of i.i.d. random variables, as in the previous example. Assuming
that the Poisson mean itself has some form of distribution also leads to a compound
Poisson distribution. The most common example of a compound Poisson distribution is
the negative binomial distribution; this is obtained by assuming a Gamma distribution for the Poisson mean i.e.

$Y|\mu \sim \text{Pois}(\mu)$ and $\mu \sim \Gamma(k,\theta)$;

then unconditionally Y has a negative binomial distribution. The negative binomial can also be obtained as a Poisson sum of the form

$$Y = X_1 + X_2 + \ldots + X_N,$$

where $N \sim \text{Pois}(\mu)$ and the X_ℓ are independent logarithmic series distribution random variables. This highlights one of the problems of compound distributions, for while they provide a more general form for the observed distribution, they are not informtive about the underlying mechanism. One of the main reasons for the use of the Gamma as a compounding distribution is that it leads to an analytically tractible compound distribution, and it also has the nice property that the means are preserved, for

$$E_y[Y] = E_\mu[E_y(Y|\mu)] = E_\mu[\mu] = \text{mean of Gamma distribution}$$

Macros are available for the negative binomial and are based on a sequence of GLIM fits, to obtain the parameter estimates for terms in the linear model for the mean, and either a Newton-Raphson or scoring method to estimate the scale parameter.

By making different assumptions about the distribution of μ it is possible to obtain a whole class of compound Poisson models. Returning to the general form of the Poisson regression model (1.1), if our aim is to account for omitted explanatory variables, a reasonable approach is to allow for the variation unexplained by the explanatory variables $X_1,\ldots,X_p$ by introducing an extra normally distributed term in the linear model, assumed to be independent of $X_1,\ldots,X_p$. Without loss of generality we can take the model to be of the following form

$$Y|X_1,\ldots,X_p,Z \sim \text{Pois}(\mu)$$
$$Z \sim N(0,1) \tag{2.1}$$

with $\log(\mu) = \beta_o + \beta_1 X_1 + \ldots + \beta_p X_p + \sigma Z$,

i.e. μ is assumed to have a lognormal distribution with location $\beta_o + \beta_1 X_1 + \ldots + \beta_p X_p$ and scale σ. This is similar to the approach suggested by Pocock et al (1981), however they apply the usual normal approximation to the Poisson distribution to obtain a final model with two sources of variation, one due to sampling and the other to unexplained variation. They show that these two components of variation can be estimated using a simple iterative procedure. However earlier remarks about approximating the Poisson distribution also apply here and for this reason we will consider a method of fitting the actual compound distribution.

Although the compound Poisson given in (2.1), which we will call the Poisson-normal compound, has no closed form it is easy to show that

$$E[Y] = e^{\underline{\beta}'\underline{x} + \frac{1}{2}\sigma^2}$$
$$\text{Var}[Y] = e^{2\underline{\beta}'\underline{x}}\left[e^{2\sigma^2} - e^{\sigma^2} \right] + e^{\underline{\beta}'\underline{x} + \frac{1}{2}\sigma^2} \tag{2.2}$$

where $\underline{\beta} = (\beta_o, \beta_1, \ldots, \beta_p)'$ and $\underline{X} = (1, X_1, \ldots, X_p)'$.

Since $e^{\sigma^2} > 1$ for $\sigma^2 > 0$ we clearly have an over-dispersed distribution. In practice we might well expect such a model to give very similar results to those obtained from the negative binomial distribution, since the Gamma and log-normal distributions can look quite similar for suitable parameter values. In the next section we will show that estimation for (2.1) is surprisingly straightforward, and more importantly the method used is easily extended to other types of compound distribution.

3. ESTIMATION FOR THE POISSON-NORMAL COMPOUND

Let us suppose that we have n observations y_i, $i = 1...n$ on some count variable and also p explanatory variables with observations $x_{1i}, \cdots, x_{pi}$, $i = 1...n$. For convenience we will introduce the following notation

$$\underline{y} = (y_1, \ldots, y_n), \quad \underline{x}_i = (x_{oi}, x_{1i}, \ldots, x_{pi})' \text{ and } \underline{X} = (\underline{x}_1, \ldots, \underline{x}_n) ,$$

where $x_{oi} = 1$, for all i. We will also denote the unobserved values of the normal compounding distribution by $\underline{z} = (z_1, \ldots, z_n)$. The compound model (2.1) implies that

$$y_i | \underline{x}_i, z_i \sim \text{Pois}(\mu) \quad \text{with} \quad \log(\mu) = \underline{\beta}'\underline{x}_i + \sigma z_i ,$$

where $\underline{\beta}$ is the vector of parameters, as above. Writing $\underline{\psi} = (\underline{\beta}', \sigma)$, the vector of unknown parameters, the joint conditional distribution of the observations is

$$f(\underline{y}|\underline{X}, \underline{z}, \underline{\psi}) = \prod_{i=1}^{n} \frac{e^{-\exp(\underline{\beta}'\underline{x}_i + \sigma z_i)}}{y_i!} \cdot \exp\{y_i(\underline{\beta}'\underline{x}_i + \sigma z_i)\} .$$

Denoting the density of the z_i by ϕ, the standard normal density, and writing

$$\phi(\underline{z}) = \prod_{i=1}^{n} \phi(z_i) ,$$

since the z_i are assumed independent,

$$f(\underline{y}, \underline{z}|\underline{X}, \underline{\psi}) = f(\underline{y}|\underline{X}, \underline{z}, \underline{\psi}) \phi(\underline{z})$$

and so

$$\log f(\underline{y}, \underline{z}|\underline{X}, \underline{\psi}) = \log f(\underline{y}|\underline{X}, \underline{z}, \underline{\psi}) + \log \phi(\underline{z})$$

$$= \sum_{i=1}^{n} \{-\exp(\underline{\beta}'\underline{x}_i + \sigma z_i) + y_i(\underline{\beta}'\underline{x}_i + \sigma z_i)\} + \log \phi(\underline{z}) + A(\underline{y}) \qquad (3.1)$$

where $A(\underline{y})$ is purely a function of $\underline{y}$.

If we had observed $\underline{z}$ estimation would proceed by the straightforward maximisation of $\log f(\underline{y}, \underline{z}|\underline{X}, \underline{\psi})$, which would correspond to fitting a Poisson regression model with $\underline{z}$ as an additional explanatory variable. However since it is unobserved a solution can be obtained by invoking the EM algorithm, see Dempster, Laird and Rubin (1977). In its most general form the EM algorithm defines an iterative procedure for the estimation of $\underline{\psi}$, as follows; the $(p+1)$th iteration involves the following two steps

E-step: Compute $Q(\underline{\psi}|\underline{\psi}^{(p)}) = E(\log f(\underline{y}, \underline{z}|\underline{X}, \underline{\psi})|\underline{y}, \underline{X}, \underline{\psi}^{(p)})$

M-step: Choose $\underline{\psi}^{(p+1)}$ to maximise $Q(\underline{\psi}|\underline{\psi}^{(p)})$

where $\underline{\psi}^{(p)}$ is the estimate of $\underline{\psi}$ from the pth iteration.

Applying this to (3.1), after some simplification the E-step becomes

$$Q(\underline{\psi}|\underline{\psi}^{(p)}) = \sum_{i=1}^{n} \left[\frac{\int_{-\infty}^{\infty} \left[-\exp(\underline{\beta}'\underline{x}_i + \sigma z) + y_i(\underline{\beta}'\underline{x}_i + \sigma z) \right] f(y_i|\underline{x}_i z, \underline{\psi}^{(p)}) \phi(z) dz}{\int_{-\infty}^{\infty} f(y_i|x_i, z, \underline{\psi}^{(p)}) \phi(z) dz} \right] \tag{3.2}$$

but unforunately this is still intractable as the integrals cannot be analytically evaluated. However this problem is easily bypassed using a numerical integration method. In particular, since the integrals are over a normal density, we can use Gaussian quadrature, see Stroud and Secrest (1966) for details and tables of quadrature points. Using k-point Gaussian quadrature methods to evaluate the integrals (3.2) becomes

$$Q(\underline{\psi}|\underline{\psi}^{(p)}) = \sum_{i=1}^{n} \left[\sum_{j=1}^{k} \frac{\left[-\exp(\underline{\beta}'\underline{x}_i + \sigma Z_j) + y_i(\underline{\beta}'\underline{x}_i + \sigma Z_j) \right] (y_i|\underline{x}_i, Z_j, \underline{\psi}^{(p)}) \pi_j}{\sum_{j=1}^{k} f(y_i|\underline{x}_i Z_j, \underline{\psi}^{(p)}) \pi_j} \right] \tag{3.3}$$

where Z_j are the quadrature points and π_j the associated weights.

Evaluating the E-step in this manner the M-step then involves setting to zero the partial derivatives of $Q(\underline{\psi}|\underline{\psi}^{(p)})$ with respect to the parameters, and solving the resulting equations, which are

$$\frac{\partial Q(\underline{\psi}|\underline{\psi}^{(p)})}{\partial \beta_r}: \sum_{i=1}^{n} \sum_{j=1}^{k} \left[-\exp(\underline{\beta}'\underline{x}_i + \sigma Z_j) + y_i \right] x_{ir} f(y_i|\underline{x}_i Z_j \underline{\psi}^{(p)}) \frac{\pi_j}{g_i^{(p)}} = 0, \quad r = 0, 1, \ldots, p$$

$$\frac{\partial Q(\underline{\psi}|\underline{\psi}^{(p)})}{\partial \sigma}: \sum_{i=1}^{n} \sum_{j=1}^{k} \left[-\exp(\underline{\beta}'\underline{x}_i + \sigma Z_j) + y_i \right] Z_j \, f(y_i|\underline{x}_i, Z_j, \underline{\psi}^{(p)}) \frac{\pi_j}{g_i^{(p)}} = 0$$

where $g_i^{(p)} = \sum_{j=1}^{k} f(y_i|\underline{x}_i, Z_j, \underline{\psi}^{(p)}) \pi_j$ is the unconditional density for y.

These equations are precisely the normal equations which would be obtained for fitting a Poisson regression model to the following data with nk observations and y-variate

$$Y = \underbrace{(y_1, \ldots, y_n, y_1, \ldots, y_n, \cdots, y_1, \ldots, y_n)}_{k \text{ copies}},$$

with explanatory variables

$$X_r = (x_{r1}, \ldots, x_{rn}, x_{r1}, \cdots, x_{rn}, \ldots, x_{r1}, \ldots, x_{rn}), \quad r = 0, 1, \ldots, p$$

$$Z = (z_1, \ldots, z_1, z_2, \ldots, z_2, \cdots, z_k, \ldots, z_k)$$

and prior weight variable $W = (w_{11}, \ldots, w_{n1}, w_{12}, \ldots, w_{n2}, \cdots, w_{1k}, \ldots, w_{nk})$

where

$$w_{ij} = f(y_i|\underline{x}_i, Z_j, \underline{\psi}^{(p)}) \frac{\pi_j}{g_i^{(p)}} \qquad i = 1, \ldots, n, \quad j = 1, \ldots, k .$$

Consequently the required iterative procedure can be easily programmed in GLIM using an iterative sequence of Poisson fits and calculations to update the weights w_{ij}.

Information about the final fit is available by comparing $-2\log$ likelihood for the fitted model with that for a complete model reproducing the data

$$D(\hat{\psi}) = -2 \sum_{i=1}^{n} \{\log(\sum_{j=1}^{k} f(y_i | \underline{x}_i, Z_j, \hat{\psi})\Pi_j) + y_i - y_i \log y_i + \log y_i!\}$$

$$= -2 \sum_{i=1}^{n} \{\log \hat{g}_i + y_i - y_i \log y_i + \log y_i!\} .$$

The important question of course is whether the compound model gives an appreciably better fit than the usual Poisson model.

Since the Poisson regression model corresponds to using a degenerate one-point distribution for the compound variable, i.e. $Z_1 = 0$, $\Pi_1 = 1$, D is comparable to the deviance for a Poisson model. The reduction in deviance, on 1 d.f., will be a measure of the importance of the inclusion of the compounding variable, although exact distributional results for this difference have yet to be determined. The importance of the compounding can also be assessed by constructing some form of interval estimate for σ and seeing if it contains zero. One simple approach is to use the maximised relative likelihood to give a profile likelihood plot for σ; details of this will be given elsewhere. The adequacy of the fit can also be verified by examining the residuals, which are easily calculated using (2.2). These can be standardised in the obvious way and adjusted residuals can also be obtained, although the calculations involved can be complicated.

4. MACROS

A set of macros have been written which enable the estimation procedure described in §3 to be used in a general situation. The user merely needs to read in the data and a set of quadrature points and then the macros can be used to set up the necessary structures, (apart from those for explanatory variables, which the user has to set up himself to allow flexibility), and to fit any specified model. Initial values for the procedure are obtained by fitting a Poisson regression model to the original data, and by default $\sigma^{(o)}$ is taken to be 1. Macro OUTPUT prints out D and also the fitted values.

The macros are listed below and their use is illustrated in §5.1.

```
! GLIM MACROS FOR POISSON-NORMAL COMPOUND
! WITH ANY SPECIFIED LINEAR MODEL IN EXPLANATORY VARIABLES
!
$MACRO SETUP !
! THIS SETS UP THE EXPANDED Y-VARIATE
! AND QUADRATURE POINTS
$CALC %M = %N*%K !
```

```
$VAR %M Y Z P W MU I J    TLP : %N HW!
$UNIT %M !
$CALC I = %GL(%N,1): J = %GL(%K,%N) !
:Y = DY(I) : Z = DZ(J) : P = DP(J) !
$USE RESET !
$$ENDM
!
$MACRO RESET !
! PROVIDES INITIAL ESTIMATES FOR FITTING A SPECIFIED MODEL
! THIS IS RE-USED WHENEVER THE MODEL IS CHANGED
$UNIT %N $ERR P $YVAR DY ! DECLARE MODEL
$CYCL $WEI
$FIT #MODL $D E$          ! FIT POISSON MODEL
$CALC TLP = %LP(I)        ! EXPAND LINEAR PREDICTOR
$UNIT %M $ERR P $YVAR Y   ! DECLARE EXTENDED MODEL
$RECY $WEI W
$CALC %I = %I + %EQ(%I,0)  ! %I = INITIAL ESTIMATE FOR Z-PARAMETER
!                         ! DEFAULT = 1
$CALC %LP = TLP + Z*%I    ! INITIAL ESTIMATES FOR LINEAR PREDICTOR
: %FV = %EXP(%LP)         ! AND FITTED VALUES
: %C = 0.01              ! DEFAULT CONVERGENCE CRITERION
: %S = 20               ! DEFAULT MAX. NO. OF ITERATIONS
: %R = 0                ! ITERATION COUNTER
$$ENDM
!
$MACRO WEI !
! THIS CALCULATES THE WEIGHT VARIATE AND
! CORRESPONDS TO THE E-STEP
$CALC MU = %EXP(%LP)             !
: W = %LOG(P) - MU + Y * %LOG(MU) !
: W = %EXP(W)                   !
: HW = 0 : HW(I) = HW(I)+W : H = HW(I)!
: W = W/H  !
$$ENDM
!
$MACR ITER
! THIS PERFORMS A SINGLE E-M ITERATION
! AND CHECKS FOR CONVERGENCE
$USE WEI  ! E-STEP
$FIT #MODL+Z $D E$ ! FIT MODIFIED MODEL
$CALC %E = -2*%CU(%LOG(HW)) ! WORKING DEVIANCE
```

```
: %T = %GE(%D-%E,%C) + %GE(%E-%D,%C) ! CHECK IF DEVIANCE HAS CONVERGED
: %D = %E  !  STORE PRESENT DEVIANCE
: %S = %IF(%LE(%T,O),O,%S-1) ! MACRO CONTROL - SET TO O IF
! DEVIANCE HAS CONVERGED OR MAX. NO. OF INTERATIONS EXCEEDED
: %R = %R+1   !  ITERATION COUNTER
$PRIN 'CURRENT WORKING DEVIANCE =' *8 %E 'NUMBER OF ITERATIONS' *-1 %R !
$$ENDM !
!
$MACRO OUTPUT
$VAR %N FV LP : %M EI
$CALC %F = -2*%CU(DY-DY*%LOG(DY)) ! CONSTANT TERM IN DEVIANCE
: %D = %E + %F  ! CORRECT DEVIANCE
$PRIN 'DEVIANCE = ' %D
$EXTR %PE
$CALC %LP = %LP - %PE(%Z)*Z  ! B'X - LINEAR PREDICTOR OF EXPLANATORY VARIABLES
: EI = %CU(1) ; EI = %IF(%GT(%EI,%N),O,EI)
: LP(EI) = %LP  ! EXTRACT LINEAR PREDICTOR B'X OF LENGTH %N
: FV = %EXP(LP+%PE(%Z)**2/2)  ! FITTED VALUES
: %LP = %LP + %PE(%Z)*Z  ! RESET %LP
$LOOK DY FV  ! OBSERVED AND EXPECTED VALUES
$$ENDM !
```

5. EXAMPLES

5.1 Fabric Data

length of roll	faults	length of roll	faults
551	6	543	8
651	4	842	9
832	17	905	23
375	9	542	9
715	14	522	6
868	8	122	1
271	5	657	9
630	7	170	4
491	7	738	9
372	7	371	14
645	6	735	17
441	8	749	10
895	28	495	7

<u>5.1 Fabric Data cont'd</u>

length of roll	faults	length of roll	faults
458	4	716	3
642	10	952	9
492	4	417	2

Table 1 Number of faults and length (ℓ) of 32 rolls of fabric

The data in Table 1 on faults in rolls of fabric might reasonably be expected to be fitted by a Poisson regression model with log ℓ as explanatory variable. However, examination of the analysis of deviance given in Table 2 shows clear evidence of a poor fit, a point which is also bourne out by examining the adjusted residuals which have mean 0.01 and variance 2.37.

	Model	Deviance	d.f.
(i)	β_0	103.7	31
(ii)	$\beta_0 + \beta_1 \log \ell$	64.6	30

Table 2 Analysis of deviance for fabric data using Poisson regression model

The compound Poisson model is easily fitted to this data using the macros given in §4 and the following GLIM commands:

```
$CALC %N = 32 : %K = 3  ! NO. OF OBSERVATIONS AND QUADRATURE POINTS
!
$VAR %N DL DY $DATA DL DY
$DINPUT I$    ! INPUT DATA FROM FILE
$CALC DL = %LOG(DL) !
$VAR %K DZ DP $DATA DZ DP $READ ! INPUT QUADRATURE POINTS
1.732 0.1667             !
0.0    0.6666            !
-1.732 0.1667           !
$MACR MODL DL$E     ! MODEL FOR INITIAL POISSON FIT
$USE SETUP          !
$CALC L=DL(I)       ! EXPAND EXPLANATORY VARIABLE
$MACR MODL L$E      ! MODEL FOR ITERATIVE FITTING
$WH %S ITER         ! FIT COMPOUND MODEL
$CALC %Z = 3        ! NO. OF Z-PARAMETER IN MODEL, FOR USE IN OUTPUT
$USE OUTPUT
```

The results from including a compounding variable in the models considered in Table 2 are summarized in Table 3, which also includes the number of iterations required to fit the models with a convergence criterion of 0.01. Comparing the fits for models (i) and (iii) shows clearly how a compounding variable can account for unexplained variation in the linear model. Model (iv) also improves the fit of model (ii) with the adjusted residuals having mean 0.03 and variance 1.03; the improvement in fit comes from the larger observations having less influence now that the variance is allowed to be greater than the mean. These results are from using a three point quadrature formula, although similar results are also obtained with larger numbers of quadrature points.

	Model	Deviance	d.f.	No. of Iterations
(iii)	$\beta_0 + \sigma Z$	64.17	30	7
(iv)	$\beta_0 + \beta_1 \log \ell + \sigma Z$	50.98	29	7

Table 3 Analysis of deviance for compound Poisson model

5.2 Quine Data

Aitkin (1978) presented an analysis of data from a sociological study of Australian Aboriginal and white children by Quine, see Table 4. The variable of interest was the number of days absent from school during the school year and the aim was to find a minimal adequate model based on four factors; age, sex, cultural background and learning ability. Aitkin considered a normal model for the response variable i.e. analysis of variance, however in the reply to the discussion he suggested the use of a log transformation, adding 1 to all values to remove the difficulties caused by zero values.

Another possible approach is to assume a Poisson distribution for the number of days absent and fit a Poisson regression model with dummy variable regression for the explanatory variables. Fitting the full model C*S*A*L gives

Table 4 Quine data on school absences.
    C = cultural background; 1 = aboriginal, 2 = white
    S = sex;   1 = female, 2 = male
    A = age group; 1 = primary, 2 = 1st form, 3 = 2nd form, 4 = 3rd form
    L = learning ability; 1 = slow, 2 = average

C	S	A	L	y		C	S	A	L	y
1	1	1	1	2 11 14		2	1	1	1	6 17 67
1	1	1	2	5 5 13 20 22		2	1	1	2	0 0 2 7 11 12
1	1	2	1	6 6 15		2	1	2	1	0 0 5 5 5 11 17
1	1	2	2	7 14		2	1	2	2	3 4
1	1	3	1	6 32 53 57		2	1	3	1	22 30 36
1	1	3	2	14 16 16 17 40 43 46		2	1	3	2	0 1 5 7 8 16 27
1	1	4	2	8 23 23 28 34 36 38		2	1	4	2	0 10 14 27 30 41 69
1	2	1	1	3		2	2	1	1	25
1	2	1	2	5 11 24 45		2	2	1	2	10 11 20 33
1	2	2	1	5 6 6 9 13 23 25 32 53 54		2	2	2	1	0 1 5 5 5 5 5 7 7 11 15
1	2	2	2	5 5 11 17 19		2	2	2	2	5 6 6 7 14 28
1	2	3	2	2		2	2	3	2	1
1	2	4	2	0 2 3 5 10 14 21 36 40		2	2	4	2	1 3 3 5 9 15 18 22 22 37
1	2	3	1	8 13 14 20 47 48 60 81		2	2	3	1	0 2 2 3 5 8 10 12 14

a deviance of 1174 on 118 d.f. showing clear over-dispersion within the cells; this is also easily seen by examining a plot of within cell variances against cell means. Fitting this model with the Poisson-normal compound distribution gives a deviance of 481 on 117 d.f., which, although a vast improvement, implies that the data is more over-dispersed than can be accounted for by this model. One possible approach is to replace the normal compounding distribution by a longer tailed distribution, this is easily done by using appropriate quadrature points in the calling of the macros, (the closed form for the expected values will no longer hold, but these can always be evaluated using numerical integration). An alternative method is to estimate the compound distribution at the same time as the model parameters, by taking a grid of quadrature points and estimating the associated weights. Model selection could in principle be carried out in the usual matter by examining a hierarchy of models, but without some form of distributional result for $D(\hat{\psi})$ it is impossible to give a formal procedure. For comparison, the minimal model given by Aitkin, S/(A+C*L), fitted using a Poisson-normal distribution, has a 'deviance' of 537 on 131 d.f..

ACKNOWLEDGEMENTS

I would like to thank my colleagues Murray Aitkin and Dorothy Anderson for their helpful comments.

REFERENCES

Aitkin, M.A. (1978). The analysis of unbalanced cross-classifications (with discussion). J.R. Statist.Soc. A, 141, 195-223.

Dempster, A.P., Laird, N.M. and Rubin, D.B. (1977). Maximum likelihood from incomplete data via the EM algorithm (with discussion). J.R. Statist. Soc.B, 39, 1 - 38.

Frome, E.L., Kutner, M.H. and Beauchamp, J.J. (1973). Regression analysis of Poisson-distributed data. J.Amer.Statist. Assoc, 68, 935-940.

Gilchrist, R. (1981). Calculation of residuals for all GLIM models. GLIM Newsletter, 4, 26 - 28.

Haberman, S.J. (1974). The analysis of frequency data. University of Chicago Press.

Nelder, J.A. and Wedderburn, R.W.M. (1972). Generalized linear models. J.R.Statist. Soc. A, 135, 370-384.

Pocock, S.J., Cook, D.G. and Beresford, S.A.A. (1981). Regression of area mortality rates on explanatory variables : What weighting is appropriate? Appl. Statist, 30, 286-296.

Stroud, A.H. and Sechrest, D. (1966). Gaussian quadrature formulas. Englewood Cliffs (N.J.) : Prentice Hall.

SOME ASPECTS OF PARAMETRIC LINK FUNCTIONS

A. SCALLAN

Polytechnic of North London

SUMMARY

In generalised linear models the mean of each observation is related to its linear
predictor via the link function. When the link function is known exactly the
maximum likelihood estimators of the parameters in the linear predictor can be
found by an iterative weighted least squares algorithm (Nelder and Wedderburn.1972).
We show how the algorithm is modified to allow for the estimation of the parameters
in models fitted with parametric link functions. Two illustrative examples are
given and some wider aspects of the method discussed.

Keywords : GLIM; PARAMETRIC LINK FUNCTIONS;
LOGISTIC CURVES; HYPERBOLIC CURVES.

1. INTRODUCTION

A generalised linear model is defined by :-

(i) Independent observations $y_1,\ldots,y_n$ arising from a distribution belonging to an
exponential family.

(ii) A set of explanatory variables $x_{k_1},\ldots,x_{kp}$ available on each observation
describing the linear predictor $n_k = \sum_i x_{ki}\beta_i$, where $\beta_1,\ldots,\beta_p$ are unknown
parameters to be estimated.

(iii) A link function $n_k = g(\mu_k)$ relating the mean of each observation, μ_k, to its
linear predictor. The inverse of the link function will be denoted by
$\mu_k = h(n_k)$.

Although such models are extremely flexible it may sometimes be necessary to
consider models which require some extensions whilst still retaining the
generalised linear model framework. As an example consider the simple logistic
curve with asymptote A:-

$$E(y_k) = A/(1 + \exp(-n_k)) \qquad ,n_k = \beta_1 + \beta_2 x_k$$

If we regard the parameter A as a known constant, and the data arise from a
distribution belonging to an exponential family, we have a generalised linear model
as specified above. However, in general, A will be unknown so we have a *parametric
link function*, as have been considered briefly by Baker *et al.*,(1980). Pregibon
(1980) indicated how the weighted least squares algorithm, used in computer packages
such as GLIM-3, might be modified to allow for the fitting of models with parametric
link functions, e.g. to find the maximum likelihood estimate of the extra parameter
A in the example above. In section 2 we describe the necessary theory and section 3
illustrates the method with two examples, including GLIM-3 coding. Some more general
aspects of the method are discussed in section 4.

2. THEORY

We consider the situation in which the inverse of the parametric link function can be written as $\mu_k = h^* (A, \eta_k)$. The log-likelihood of an individual observation is given by

$$\log(P(y_k)) = (y_k \, \theta_k - b(\theta_k))/a_k(\phi) + c(y_k, a_k(\phi)),$$

with the properties : $E(y_k) = b'(\theta) = \mu_k,$

$$\text{var}(y_k) = a_k(\phi). \; b''(\theta_k) = \sigma_k^2.$$

To derive the maximum likelihood estimates for $\beta_1, \ldots, \beta_p$, A we need to solve the likelihood equations $\partial L/\partial \beta_1 = \ldots = \partial L/\partial \beta_p = \partial L/\partial A = 0$, where L is the joint log-likelihood of the observations.

Now, $\partial L/\partial \beta_i = \sum_k \partial l_k/\partial \beta_i$

$$= \sum_k (\partial l_k/\partial \theta_k).(\partial \theta_k/\partial \mu_k).(\partial \mu_k/\partial \eta_k).(\partial \eta_k/\partial \beta_i) \; ,$$

$$= \sum_k (y_k - \mu_k) \delta_k f_k. \; x_{ki} \; , \qquad (2.1)$$

where $\delta_k = 1/(\partial \mu_k/\partial \eta_k)$ and $f_k = 1/(\alpha_k(\phi).b''(\theta_k).\delta_k^2)$.

similarly, $\dfrac{\partial L}{\partial A} = \sum_k (\partial l_k/\partial \theta_k).(\partial \theta_k/\partial \mu_k)(\partial \mu_k/\partial A) \; ,$

$$= \sum_k (y_k - \mu_k).f_k.\delta_k \, m_k \; , \quad (2.2)$$

where $m_k = (\partial \mu_k/\partial A). \; \delta_k.$

Thus, by defining a new explanatory variable $\underline{m}$, we can write the likelihood equation for A (2.1) in exactly the same form as the equations for estimating the β's (2.2). The maximum likelihood estimates of all the parameters are then found by the usual weighted least squares algorithm, but with the inclusion of an extra explanatory variable, $\underline{m}$, in the linear predictor to allow for the estimation of A.
Since $\underline{m}$ is a function of the iterative weights and fitted values it must be recalculated at each iteration. In the next section we show how this type of model can be fitted quite easily using GLIM.

3. EXAMPLES

3.1 *Logistic and Hyperbolic Curves.*

Consider the fitting of a hyperbolic curve with an unknown origin of the form:-

$$E(y_k) = \mu_k = (x_k + A)/(\beta_1 + \beta_2 x_k)$$

$$= h^* (A, \eta_k).$$

The quantities we used in order to solve the least squares equations are δ_k, f_k and m_k.

Now, $\delta_k = \partial \eta_k/\partial \mu_k = - (x_k + A)/\mu_k^2 \; ,$

and $f_k = 1/(b''(\theta_k). \delta_k^2)$

$$= 1/(\mu_k^2 \cdot \delta_k^2) = \mu_k^2/(x_k + A)^2 \ ,$$

- assuming Gamma errors.

Also, $\quad \partial \mu_k/\partial A = 1/\eta_k = \mu_k/(x_k + A) \quad ,$

hence $\quad m_k = (\partial \mu_k/\partial A) \cdot \delta_k = - 1/\mu_k \ .$

As a second example we consider the fitting of a generalisation of the logistic curve described earlier. This can be written as:-

$$E(y_k) = A/(1 + \exp(-\eta_k/B)^B$$

The power parameter, B, cannot easily be estimated by the method we have described because of problems with its stability. These will be discussed in section 4. However, we can estimate this parameter by minimising the residual sum of squares for different values of B. We now show how the extra parameter A can be estimated for a fixed value of B.

In many situations where a logistic curve is fitted the data are assumed to follow a lognormal distribution (Nelder,1961). Hence by taking logarithms of the observations we arrive at the model:-

$$\log(y_k) = \log(A) - B.\log(1+\exp(-\eta_k/B) + \varepsilon_k$$

$\quad$ where ε_k is distributed as $N(0,\sigma^2)$

Hence $E(y_k) = \mu_k = h*(A,\eta_k).$

Now, $\quad \delta_k = 1/(\partial \mu_k/\partial \eta_k).$

$\quad = (1 + \exp(-\eta_k/B))/\exp(-\eta_k/B)$

and $f_k = 1/(b''(\Theta_k) \cdot \delta_k^2) = 1/\delta_k^2$, assuming normal errors.

Also, $\partial \mu_k/\partial A = 1/A$, so that

$\quad m_k = \delta_k/A \ .$

3.2 GLIM *Coding*

The models can be fitted using the OWN facility of GLIM, although with a few modifications since we need to recalculate the iterative explanatory variable and incorporate it into the linear predictor at each iteration. The following macros for the logistic curve are self explanatory if it is noted that %LP represents the modified linear predictor and LP the original linear predictor, $\underline{\eta}$. See Baker and Nelder (1978) for details of the language etc.

```
$MAC FV
$C Macro INIT provides starting values, MEXT updates A and the linear predictor.
$CALC %S = %NE(%PL,0) $SWITCH %S  INIT  MEXT
$USE DR $CALC M=(1/%A) * %DR
$CALC %FV = %LOG(%A)- %B*%LOG(1+%EXP(-LP/%B)) : %LP=LP+%A*M
$END
```

```
$MAC DR $CALC %DR = (1+%EXP(-LP/%B))/%EXP(-LP/%B)   $END
$MAC VA $CALC %VA = 1  $END
$MAC DI $CALC  %DI = (%FV- %YV)**2   $END
$MAC INIT $CALC LP = -%B*%LOG(%EXP((%LOG(%A) - %YV)/%B)-1)   $END
$MAC MEXT $EXTRACT %PE $CALC %A = %PE(3) : LP = %LP-%A*M    $END
```
GLIM output for this model is given in the appendix.

4. DISCUSSION

4.1 *Efficiency and Stability.*

The method described above has been found to be reasonably efficient for the models
considered in section 3, convergence being usually obtained within four iterations.
However, the efficiency will be dependent both on the type of model fitted and the
accuracy of the starting values.

By setting the initial fitted values to the observed values we can restrict to A
alone the number of parameters for which we need provide an initial estimate.
Sometimes the value of A will be know fairly accurately; for example when A is the
asymptote in the logistic curve it can often be estimated accurately from the raw
data. However, in many cases the initial estimate might be far from the 'true' value
and this can, though may not always, lead to instability of the process. For
example, while giving a 'wild' initial estimate for the asymptote in the logistic
curve has caused the algorithm to break down, it seems able to cope with almost any
initial estimate for the origin of the hyperbolic curve.

This potential instability is well demonstrated by a model which is an obvious
candidate for analysis using this method, namely the power link $\eta = \mu^A$. The iterative
explanatory variable is easily seen to be $m = - \mu^A.\log(\mu)$ and in all other respects
this model is of the same type as those considered previously. However, attempts
at fitting this model to both simulated and experimental data have shown that small
deviations in the starting value of A from its 'true' value causes the algorithm to
break down.

4.2. *Extensions*

We first note that any function of the extra parameter A, can be estimated in its
place. For example suppose $\alpha = \alpha(A)$, then $\partial L/\partial \alpha = (\partial L/\partial A).(\partial A/\partial \alpha)$ and the iterative
explanatory variable for estimating α is given by $m^1 = m.(\partial A/\partial \alpha)$. Thus, in the case
of the logistic curve, we could estimate A or, as is more usual, $\alpha = \log(A)$.   In
some cases a reparameterisation of this form can improve the estimation procedure,
e.g. by making the likelihood concave and, hence, the maximum more well-defined.
For a further discussion of this see Ross, 1980.

Secondly, there is no reason why the method should not, in theory, work for models with several extra parameters; i.e. with link functions of the form $\mu=h^*(A_1,\ldots,A_j,\eta)$. The weighted least squares algorithm is modified in the same way as for one parameter. Thus each A_i is estimated by fitting an extra explanatory variable m_i, where $m_i = (\partial\mu/\partial A_j).\delta$. We might, for example, consider fitting hyperbolic curves of the form

$$\mu_k = \overline{\prod_{i=1}^{r}} (x_{ki} + A_i)/\eta_k ,$$

where η_k is a polynominal in the r variables x_i. Here the extra explanatory variable for estimating the ith origin, A_i, is given by:-

$$m_{ki} = -\overline{\prod_{j\neq i}} (x_{kj} + A_j)/\mu_k.$$

ACKNOWLEDGEMENTS

I would like to thank Dr.R.Gilchrist and Dr.M.Green, Polytechnic of North London for continued help and encouragement and Mr.R.Baker, Rothampstead Experimental Research Station, for many helpful comments. I would also like to thank a referee who commented on an earlier version of the paper.

REFERENCES

BAKER, R.J. and NELDER, J.A. (1978). *The GLIM system, Release 3*. Oxford: Numerical Algorithms Group.

BAKER, R.J. , PIERCE, C.B. and PIERCE, J.M. (1980). Wadley's problem with controls. The GLIM *Newsletter*. Dec.1980, 32-35

BOX, G.E.P.and TIDWELL,P.W.(1962). Transformation of the Independent Variables. *Technometrics*. Vol.4,No.4,531-550.

NELDER, J.A.(1961). The fitting of a generalisation of the logistic curve. *Biometrics*, 17, 89-100.

NELDER, J.A.and WEDDERBURN,R.W.M.(1972). Generalised linear models. *J.R.Statist.Soc A*, 135, 370-383.

PREGIBON, D.(1980). Goodness of link tests for generalised linear models. *Appl.statist*.29, No.1,15-24.

ROSS,G.J.S.(1980). Use of non-linear transformation in non-linear optimisation problems. *COMPSTAT 1980*, 382-388.

APPENDIX

```
$CALC Y=%LOG(Y)     \C Take Logs as a Normalising transformation.

$OWN FV DR VA DI    \C Declare OWN model and dependent variable.
$YVAR Y
$CALC %B=1.0:%A=75  \C Choose value for B and initial estimate for A (which
                       must be greater than the largest observed value).
$CALC %LP=M=0       \C GLIM requires initial values for the linear predictor
                       and we must assign some initial value to M.

$FIT X+M$           \C Fit original model + the extra explanatory variable, M.


CYCLE  DEVIANCE     DF
  3    0.6376E-02    8


$DISP EV            \C Display parameter estimates and covariance matrix.

       ESTIMATE     S.E.        PARAMETER
  1     -1.130      0.3572E-01  %GM
  2      0.8542     0.1462E-01  X
  3     73.26       1.348       M
  SCALE PARAMETER TAKEN AS   0.7970E-03

(CO)VARIANCE MATRIX
  1   1.2762E-03
  2   3.8103E-04  2.1362E-04
  3  -4.2598E-02 -1.1913E-02   1.817
        1            2           3
  SCALE PARAMETER TAKEN AS   0.7970E-03


$LOOK X Y %FV       \C Look at fitted and observed values,remembering
                       we have already taken Logs.

         X           Y           %FV
  1    -2.150       1.273       1.278
  2    -1.500       1.833       1.797
  3    -0.8500      2.255       2.293
  4    -0.8000E-01  2.828       2.832
  5     0.5200      3.199       3.200
  6     1.100       3.520       3.501
  7     2.280       3.912       3.928
  8     3.230       4.128       4.115
  9     4.000       4.239       4.197
 10     4.650       4.206       4.237
 11     5.000       4.239       4.252
```

A MODEL FOR A BINARY RESPONSE WITH MISCLASSIFICATIONS

Anders Ekholm and Juni Palmgren

Department of Statistics, University of Helsinki

SF-00100 Helsinki, Finland

SUMMARY

Observations on a binary response may be subject to misclassification. A linear logit model for the true binary response is specified and estimated jointly with the error probabilities for the two types of misclassification. The model is illustrated using a subset of the well-known coal-miners data. The problem is formulated as an incomplete data problem and the EM-algorithm and GLIM is used for estimation. The connection to latent structure models is discussed.

Keywords: Errors of observation; Incomplete data; Latent variable; Logit model

1. INTRODUCTION

Consider the data of Table 1 collapsed from a set given by Ashford and Sowden (1970). The most important information therein is the relative frequency of wheeze at different ages. The obvious way of summarizing these data is to fit a logit (or probit) model. The fit for a logit model is quite reasonable. The value of the deviance is 8.2 on 7 degrees of freedom.

TABLE 1

Number of coalminers examined and number suffering from wheeze

Age-group in years	Midpoint x	Number of wheeze	Number examined	Relative frequency
20–24	22.5	104	1952	0.05
25–29	27.5	128	1791	0.07
30–34	32.5	231	2113	0.11
35–39	37.5	378	2783	0.14
40–44	42.5	442	2274	0.19
45–49	47.5	593	2393	0.25
50–54	52,5	649	2090	0.31
55–59	57.5	631	1750	0.36
60–64	62.5	504	1136	0.44

It is, however, conceivable that wheeze cannot be diagnosed without errors. We suppose that errors of diagnosis, or more generally of observation, go both ways. Some individuals are classified as suffering from wheeze though they do not, and some cases of wheeze pass unnoticed. The probabilities of these different types of errors are not necessarily equal, and neither probability is supposed to be tied to the age of the patient.

We consider thus the following model. The latent binary response variable η (true wheeze) follows a linear logit model with x (age) as a fixed explanatory variable, measured without errors:

$$pr(\eta(x)=1|x) = \pi(x) = \exp(\alpha+\beta x)/(1 + \exp(\alpha+\beta x)). \tag{1}$$

The manifest binary response variable Y (diagnosed wheeze) relates to the latent response subject to the errors of observation in the following way: for a fixed x

$$pr(Y(x)=1|\eta(x)=0) = \epsilon_0 \tag{2}$$

$$pr(Y(x)=0|\eta(x)=1) = \epsilon_1.$$

The values used for x in the above equations are the midpoints of the age-intervals in Table 1. We do not have access to the original data with age for each person. The model applies, however, in principle equally to the case where the data are not grouped.

The probabilities of the 2 x 2 table of the latent and the manifest response are according to (1) and (2)

Latent (η)	Manifest (Y) 1	0
1	$\pi(x)(1-\epsilon_1)$	$\pi(x)\epsilon_1$
0	$(1-\pi(x))\epsilon_0$	$(1-\pi(x))(1-\epsilon_0)$

$$\tag{3}$$

The probabilities of the latent margin are, in accordance with (1), simple, but the probabilities of the manifest margin are mixtures of the probability of an error of one kind and the probability of no error of the other kind. This makes the likelihood complicated.

We shall not attempt a full appraisal of the identifiability of the parameters ϵ_0, ϵ_1, α and β from the manifest responses, but want to make the following points:

(i) There is no information, whatsoever, about the parameters α and β in the data, if $\epsilon_0 = 1 - \epsilon_1$. This is an academic point, since it would anyway be unreasonable to use this model if the probability of either kind was greater than 1/2. We expect the relevant range of the error probabilities to be from zero to roughly 1/3.

(ii) The probability of a "wrong" manifest response, that is $pr(Y \neq \eta)$, is independent of the latent response if and only if $\epsilon_0 = \epsilon_1$. This special case has no particular bearing on the identifiability of the parameters, except when $\epsilon_0 = \epsilon_1 = 1/2$, which is a special case of (i).

(iii) Consider the 2 x 2 table (3) when $x \to \infty$ and (1) is valid. The probability of a zero manifest response is then a function only of ϵ_1. Analogously, when $x \to -\infty$ the probability of a unit reponse is a function only of ϵ_0. It is thus clear that for perfect data over a wide enough range of x-values the parameters ϵ_0, ϵ_1 and accordingly also α and β are identifiable.

(iv) It is not clear whether there exist cases other than the one mentioned in (i), where two or more points of the space of the four parameters ϵ_0, ϵ_1, α, β are mapped on to the same manifest marginal probabilities in (3). The argument in (iii) shows that this would have to be when the range of the x-values is restricted.

It is, however, clear that questions of identifiability aside the problems of finding estimates for the parameters from less than perfect data are difficult if the range of x-values available is restricted to the part where $\pi(x)$ is almost linear. Before turning to the problems of estimation we find it useful to formulate this model as a log-linear one, and to explore the connection with the latent class models of Haberman (1979) and Goodman (1974).

2. THE LOG-LINEAR MODEL

We assume that the explanatory variable x takes r distinct values and introduce a categorical variable T with r levels, one for each distinct value of x.

We now reparametrize the model defined in (1) and (2) as a log-linear model for the 2 x 2 x r frequency table formed by the latent variable η and the two manifest variables Y and T.

Let Z_{ijk} denote the number of observations for which η=i, Y=j and T=k (i,j=0,1; k=1,..,r). Now, since the distribution over the two categories of η cannot be observed we cannot observe the complete data Z_{ijk} but only the incomplete data $U_{jk}=\Sigma_i Z_{ijk}$.

We specify, however, the log-linear model of conditional independence

$$\log E(Z_{ijk}) = \tau_k + \gamma_i + \omega_j + \rho_{ij} + \delta_i x_k \tag{4}$$

for the unobserved complete data. We adopt the GLIM convention of restricting the parameters by setting a parameter with any of its indices at its first value equal to zero. Thus we may drop the indices i and j from the parameters in model (4). Now, γ, ω and τ_k (k=1,..,r) represent the main effect parameters for η, Y and T respectively and ρ represents the interaction between η and Y. The interaction between η and T is specified as a linear function of the x-values with slope parameter δ.

We get the following one-to-one relation between on une hand the original parameters ϵ_0, ϵ_1, α and β of model (1)-(2) and on the other hand the parameters γ, ω, ρ and δ in the log-linear model (4): $\epsilon_0 = 1/\exp(\omega+\rho)$, $\epsilon_1 = \exp(\omega)/(1+\exp(\omega))$, $\alpha = \ln((1+\exp(\omega)/(1+\exp(\omega+\rho))-\gamma$ and β=-δ. Note, that when γ, ω, ρ and δ are known the values of the parameters τ_1, τ_2,.., τ_r are fixed through the number of observations in each category of T.

In GLIM notation we can write model (4) in the form

T + η + Y + η.Y + η.X

We now refer to the latent class models of Haberman through his example on abortion attitudes (Haberman, 1979, p.542). In his example B,D and F denote three binary variables representing "yes" or "no" answers to three questions on abortion. Surveys were made in 1972, 1973 and 1974, thus T denotes the year of survey and has three levels. Haberman assumes the manifest variables B,D and F to be conditionally independent given the level of a binary latent variable η. He specifies a log linear model

$$T + η + B + D + F + η.B + η.D + η.F + η.T$$

for the incompletely observed η x B x D x F x T frequency table.

Note, that the purpose of Haberman's latent class model and our model involving errors of observation in a binary reaponse are quite different. Haberman explains the interaction structure of the answers B, D and F through the latent classes of η representing a more general underlying conservative versus liberal attitude. In our model emphasis is put on the structural part describing the dependence of η on x.

Haberman's model on abortion attitudes can be illustrated through the diagram in Fig. 1a and a corresponding diagram of our model is shown in Fig. 1b.

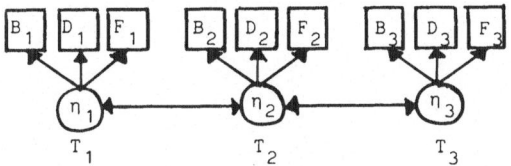

Fig.1a   Haberman's model on abortion attitudes

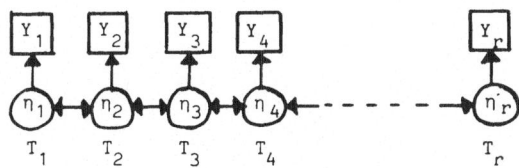

F ig.1b   A logit model with errors of observation

In Haberman's model the relation between the three manifest indicators is structured by assuming conditional independence given the latent variable. The relation between the manifest indicators and the latent variable is constant over time. No structure is imposed on the development of η over time. In our model there is only one manifest indicator for each latent variable and their relation is the same for each T. A very strong structure is imposed on the relation between η_1, η_2, ..., η_r.

Although the interpretation of the two models are different they can be specified as equivalent log-linear models of conditional independence for an incompletely observed frequency table.

3. ESTIMATION

We restrict ourselves to maximum likelihood estimation. In each category of the explanatory variable the total number of observations is treated as fixed. For each k (k=1,..,r) the frequencies Z_{ijk} are multinomially distributed and thus the observed marginal frequencies U_{jk} are binomial.

We are dealing with incomplete data from an exponential family and to obtain maximum likelihood estimates of the unknown parameters we could choose between the following procedures:

(i) straight forward maximisation of the log likelihood function.

(ii) an iterative weighted least squares procedure, based on Fisher's scoring algorithm (Haberman, 1979). The attractive feature of this algorithm is that the asymptotic covariance matrix for the parameters is produced directly.

(iii) The EM-algorithm (Dempster, Laird and Rubin, 1977).

The estimates in Section 4 were computed through alternative (i) using a NAG library routine and through alternative (iii) using GLIM. After obtaining the estimated expected cell frequencies for the complete data the asymptotic covariance matrix for the parameters in the log-linear model was computed via the formula derived in (ii).

Although the use of the EM-algorithm in GLIM is too slow to be really useful, the procedure is briefly described in Section 3.1.

3.1. The EM-algorithm and GLIM

We briefly describe the steps involved in using GLIM to program the EM-algorithm for the example in Section 1.

(a) Read the observed frequencies U_{jk} of the Y x T margin.

(b) Give initial values to the error probabilities ε_0 and ε_1 defined in (2).

(c) Compute initial values Z_{ijk} for the complete data.

(d) Use the following MACRO iteratively until convergence:

Maximisation step: Maximize the log likelihood for the estimated complete data assuming the log-linear model (4). The crucial GLIM statements are \$YVAR Z \$ERROR P \$LINK L \$FIT -%GM + T + η + Y + η.Y + η.X. We make use of the well known connection between the Poisson and the multinomial distributions.

Expectation step: Compute the estimated conditional expected frequencies for the complete data, given the observed Y x T margin $Z_{ijk} = U_{jk} \quad m_{ijk} / m_{.jk}$. Here m_{ijk} are the fitted values from the preceding maximisation step and the dot denotes summation over the index. Return to the maximisation step.

(e) Note, that the deviance and the asymptotic covariance matrix for the parameters produced in GLIM are not valid when the data are incomplete. Compute the deviance D using the formula $D = 2 \Sigma_{jk} U_{jk} \ln(U_{jk}/ \hat{m}_{.jk})$ where $\hat{m}_{.jk}$ are the maximum liklihood estimates obtained in (d). Compute the asymptotic covariance matrix for the parameters using the procedure of Section 3.2.

3.2. The covariance matrix

When presenting the scoring algorithm Haberman (1979) shows that the asymptotic covariance matrix for the p = r + 4 parameters in model (4) can be put into the well-known form $(A^T WA)^{-1}$. The A matrix has dimensions (2r x p) and elements a_{ts} = $(\Sigma_i c_{its} m_{it})/ \Sigma_i m_{it}$ (t=1,..,2r; s=1,...,p), where t is an index running through the levels of the joint Y x T margin. The constants c_{its} are elements in the design matrix for model (4).

One could say that the matrix A corresponds to a derived "design" matrix for the incomplete data model, whose elements are weighted sums of the elements in the design matrix for the corresponding complete data problem. The matrix W is a diagonal (2r x 2r) matrix with elements m_t.

Note, that the covariance matrix computed from the Poisson based likelihood function gives the correct submatrix for the parameters of interest (Palmgren, 1981). The asymptotic standard errors for the original parameter estimates $\hat{\epsilon}_0$, $\hat{\epsilon}_1$, $\hat{\alpha}$ and $\hat{\beta}$ are derived from the above covariance matrix using standard first order approximation formulas.

4. EXAMPLE

We have fitted the model presented in Section 1 to the data of Table 1. The value
for the deviance is 3.73 on 5 degrees of freedom. The deviance difference between this
model and the ordinary logit model is 4.5 on 2 degrees of freedom. As is seen from
Fig.2 the linear logit model specified without errors and our model fit the data equally well. The estimates of the parameters and their standard errors are given in Table
2, together with the corresponding values for the intercept and slope of the logit
model without errors.

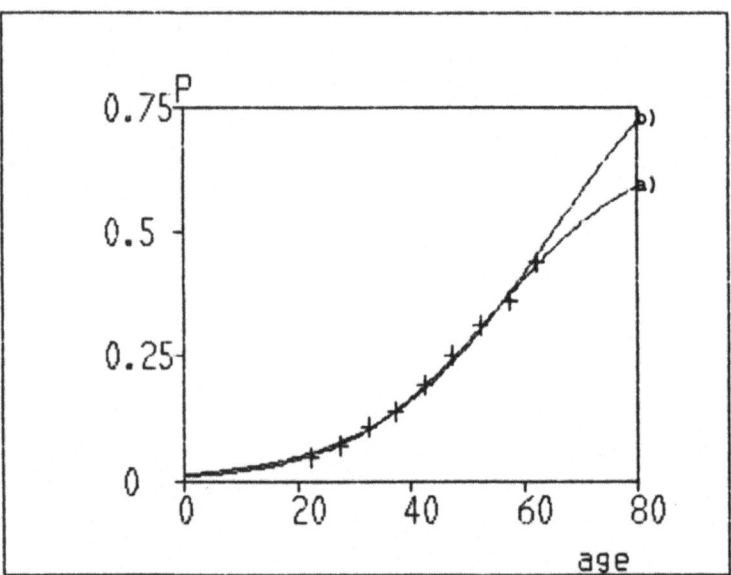

Fig.2 Observed relative frequencies and estimated probabilities of
diagnosed wheeze according to a) the misclassification model
and b) the ordinary logit model.

TABLE 2

The estimates of the parameters and their standard errors of the
ordinary logit model and the logit model with errors of observation

Logit model	0.000530	0.322	-4.200	0.0765
with errors	(0.0168)	(0.131)	(0.151)	(0.0129)
Ordinary	-	-	-4.258	0.0652
logit model	-	-	(0.0847)	(0.00177)

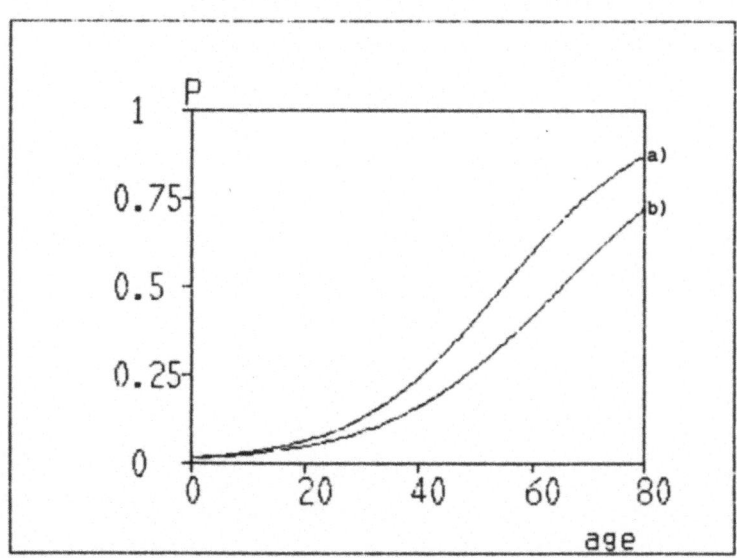

Fig.3 The probabilities for a) true wheeze and b) diagnosed wheeze
as functions of age.

The most interesting feature of the model where errors of diagnosis are presumed is the fact that the estimate for diagnosing wheeze when the person is well is negligible. The percentage of unnoticed wheeze is, on the other hand, very high. Almost a third of the cases go unnoticed. The curve of our model in Fig.2 coincides with the curve of the ordinary logit model at x=0, but the upper asymptotes differ markedly. This is a reflection of the fact that $\hat{\epsilon}_o$ is negligible but $\hat{\epsilon}_1$ considerable. The referee points out that it would be sensible to refit the model with $\epsilon_o=0$. This might, in fact, concentrate the deviance difference of 4.5 in 1 degree of freedom.

In Fig.3 the probabilities for "true" and diagnosed wheeze are plotted. The curve for true wheeze is computed according to equation (1). Note from Table 2 that the slope parameter for true wheeze is 17% higher than for diagnosed wheeze. This means, for example, that the estimated probability of true wheeze for the oldest age-group is 0.64, as compared to the observed relative frequency 0.44.

The weak point of this model is that the interplay between the latent and the manifest responses might, in fact, be much more complicated than we have assumed. The statistical assessment would have to be supplemented by a subject matter assessment. It certainly is essential that the data concern self-reported wheeze.

5. DISCUSSION

The theory of latent structure analysis has been explored and explained thoroughly by statisticians lately. Some important references are Andersen (1982) (referred to below as A), Bartholomew (1980) (B), Bock and Aitkin (1981) (B&A) and Muthén (1981) (M). The most elaborate exploitation of the theory of latent variables seems to be the LISREL-theory developed by Jöreskog and Sörbom (1979) (J&S). One way of structuring the vast and diversified field would be to cross-classify the models presented by two dichotomous criteria: (i) Does the model use quantitative or qualitative latent variables. (ii) Does the model assume a structural relation between a set of latent response variables and a set of latent explanatory variables, or just a relation between a set of latent variables and a set of manifest ones.

We tabulate below the four ensuing cases as a 2 x 2 table and indicate where the models referred to above belong:

		The latent variables are	
		qualitative	quantitative
The latent variables are	response and explanatory	G	J&S, M
	just explanatory	A, H	A, B, B&A

G stands for models treated by Goodman, H for Haberman and the other letter combinations are indicated above. When one assumes a relation between two sets of quantitative variables the obvious way of doing this is to work with a set of regression equations. If one wants the regression equations to contain all information about the relations one has to supplement them with an assumption of a probability model.

It seems to us that one of the most appealing features of the theory of generalized linear models for Poisson and multinomial data is that one can explore the structure of the full simultaneous distributions of several qualitative variables. One is not

limited to summarizing the relation between two sets of variables in one universally
valid set of equations. In stead, one can work with a relation between for instance
two variables that is structurally different for different levels of a third qualitative factor. We believe that many relations in social sciences are best approached as
potenttially very different for different subgroups. Sex, race education and many other social conditions might be qualitative conditioning factors of this kind.

This remark combined with the cross-classification above amounts to the point that
it might be worth while to model some problems in social science as relations among
latent qualitative variables. One would then estimate the values of the relevant interaction terms from qualitative manifest variables. The latent variables would be necessary, for instance, because the manifest categories are contaminated by misclassi-
fications.

This view prompted our interest in the model presented here. The model does not
fulfil the criteria for being in the north-west cell, for the structural relation assumed is between a fixed manifest explanatory variable and a latent response. But both
the latent and the manifest response variables are qualitative.

REFERENCES

ANDERSEN, E. B. (1982). Latent structure analysis: A survey. Scand. J. Statist. 9,1-12.

ASHFORD, J. R. and SOWDEN, R. D. (1970). Multivariate probit analysis. Biometrics 26, 535-46.

BARTHOLOMEW, D. J. (1980). Factor analysis for categorical data. J. Roy. Statist. Soc. B. 42, 293-321.

BOCK, R. D. and AITKEN, M. (1981). Marginal maximum likelihood estimation of item parameters: Application of an EM algorithm. Psychometrika, 46, 443-59.

GOODMAN, L. A. (1974). The analysis of systems of qualitative variables when some of the variables are unobservable. Part I: A modified latent structure approach. Amer. J. Social. 79, 281-361.

HABERMAN, S. J. (1979). Analysis of qualitative data. Vol. 2. Academic Press, N. Y.

JÖRESKOG, K. G. and SÖRBOM, D. (1979). Advances in factor analysis and structural equation models. Abt Books, Cambridge, Massachusetts.

MUTHEN, B. (1981). A general structural equation model with ordered categorical and continuous latent variable indicators. Reserch report 81-9. University of Uppsala, Sweden.

PALMGREN, J. (1981). The Fisher information matrix for log-linear models arguing conditionally on observed explanatory variables. Biometrika, 68,563-555.

LOGLINEAR MODELS WITH COMPOSITE LINK FUNCTIONS IN GENETICS

R. BURN

Department of Mathematics
Brighton Polytechnic

SUMMARY

Loglinear models with linear composite link functions constitute a very flexible class
of models which can accommodate a wide range of estimation problems for probability
models in genetics. It is shown how GLIM-3 may be used to obtain estimates of gene
frequencies, inbreeding coefficients and other parameters and their asymptotic covariances and standard errors. Examples are taken from medical genetics and evolutionary
biology.

KEYWORDS: INDIRECT OBSERVATION; COMPOSITE LINK FUNCTIONS; BLOOD GROUP DATA;
INBREEDING; CRYPTIC GENOTYPES; TRISOMY.

1. INTRODUCTION

A feature of many statistical problems in genetics is an underlying frequency table
which can be observed only indirectly. To be specific, the data will usually consist
of phenotype counts and these counts are the sums of frequencies of various subsets
of the cells of an unobservable table of genotype frequencies. In most (but not all)
of the problems considered here, this situation arises because of the phenomenon of
dominance. This means that certain genotypes are indistinguishable from certain others
so that, in general, by observing phenotype counts, we are effectively observing the
genotype frequency table but with some of the cells grouped together.

An example is the human ABO blood group system in which there are three alleles, A, B
and O, and the dominance relations are such that there are only four observable phenotypes A, B, AB and O. Blood group A consists of genotypes AA and AO, B consists of BB
and BO, blood group AB corresponds to genotype AB and group O to genotype OO. The
problem is to obtain estimates of the gene frequencies (probabilities) p, q and r of
A, B and O, respectively, from the phenotype counts n_A, n_B, n_{AB} and n_O in a sample of
N individuals. Following Thompson and Baker (1981), we can regard the underlying
frequency table as a 3 x 3 contingency table in which rows and columns correspond to
the gene contributed to the individual by each of its parents. If we assume random
mating and that the population is in Hardy-Weinberg equilibrium (Elandt-Johnson (1971)),
then the expected cell frequencies are as shown in Table 1.1. Genetic hypotheses other
than random mating could be considered and would lead to different expected frequencies.
The GLIM approach to this problem, suggested by Thomson and Baker (1981), not only
leads to estimates of p, q and r, but also provides a test, by means of the deviance,

TABLE 1.1

	A	0	B
A	Np^2	Npr	Npq
0	Npr	Nr^2	Nqr
B	Npq	Nqr	Nq^2

of the genetic hypothesis from which the expected frequencies were obtained. An example of this is presented in Section 4 in which the random-mating hypothesis is rejected in favour of a model with inbreeding.

Maximum likelihood estimation of gene frequencies with indirect observation has been explored, along with other applications, by Haberman (1974, 1977). The recent article by Thompson and Baker (1981) set the ABO estimation problem in the framework of a generalised linear model with a composite link function and showed how GLIM-3 can be persuaded to fit these models. Our aim is to extend the method of Thompson and Baker to a number of other estimation problems in genetics. These include the estimation of gene frequencies in a two-locus model in medical genetics and a problem in evolutionary genetics which involves a slightly more subtle version of dominance. (Incidentally, this last problem amounts to a case of estimating the mixing probabilities in a finite mixture with known kernel and suggests the possibility of using GLIM for this general class of problems.) These applications are considered in Sections 3 to 6 after reviewing the theoretical background in Section 2.

The computations were performed using GLIM-3 and listings are available on request from the author. At the time of writing, GLIM-4 is not yet available, but it promises to simplify the programming considerably when it is released.

2. THEORY

If the genotype frequencies were directly observable then we could set up an ordinary loglinear model relating expected cell frequencies to the parameters. For instance, in the ABO problem, let $\underline{y}$ denote the 9 x 1 vector of expected frequencies and let $\underline{\beta} = (\ln N, \ln p, \ln r, \ln q)^T$. Then the random mating model is

$$\ln (\underline{y}) = \underline{\eta} \ , \qquad \underline{\eta} = X\underline{\beta}$$

where

$$X^T = \begin{bmatrix} 1 & 1 & 1 & 1 & 1 & 1 & 1 & 1 & 1 \\ 2 & 1 & 1 & 1 & 0 & 0 & 1 & 0 & 0 \\ 0 & 1 & 0 & 1 & 2 & 1 & 0 & 1 & 0 \\ 0 & 0 & 1 & 0 & 0 & 1 & 1 & 1 & 2 \end{bmatrix}$$

The expected frequencies corresponding to the available data, however, are
$\underline{\mu} = (\mu_A , \mu_B , \mu_{AB} , \mu_0)^T$, and this is related to $\underline{y}$ by $\underline{\mu} = C\underline{y}$, where

$$C = \begin{bmatrix} 1 & 1 & 0 & 1 & 0 & 0 & 0 & 0 & 0 \\ 0 & 0 & 0 & 0 & 0 & 1 & 0 & 1 & 1 \\ 0 & 0 & 1 & 0 & 0 & 0 & 1 & 0 & 0 \\ 0 & 0 & 0 & 0 & 1 & 0 & 0 & 0 & 0 \end{bmatrix} .$$

In general, suppose we have data $\underline{y} = (y_1, \ldots, y_n)^T$ and let $\underline{\mu} = E(y)$. We assume that the μ_i are linear combinations of expected cell frequencies γ_j , $j = 1, \ldots, m$: $\underline{\mu} = C\underline{\gamma}$. The γ_j are related to parameters $\beta_1, \ldots, \beta_p$ through a loglinear model : $\underline{\gamma} = \exp(\underline{n})$, $\underline{n} = X\underline{\beta}$ where X is a known m x p matrix corresponding to the hypothesis being considered. This is an instance of a generalised linear model with composite link function, introduced by Thompson and Baker (1981). The models considered in that paper are more general than ours in that (a) they have a general link $\underline{\gamma} = h(\underline{n})$, and (b) the relationship between $\underline{\mu}$ and $\underline{\gamma}$ is allowed to be non-linear : $\mu_i = c_i(\underline{\gamma})$. However, loglinear models with linear composite link functions will be sufficient for our purposes. For these models, Thompson and Baker show that by taking $\underline{z} = CH\underline{n} + (y - \underline{\mu})$ as working dependent variable, and X* = CHX as working independent variables, where $H = \text{diag} \{\gamma_1, \ldots, \gamma_m\}$, GLIM's $OWN directive may be used to obtain maximum likelihood estimates. As initial values for $\underline{\beta}$ it has been found convenient to use $\beta_j = \frac{1}{2} \ell n(N/m)$, where N is the sample size and m the number of cells in the underlying frequency table.

We choose a parameterisation for which X has rank p - 1 and the first element of $\underline{\beta}$ is ℓn N and, in order to avoid intrinsic aliasing, we fit a model without the constant term (%GM), thus removing the parameter ℓn N. Assuming the remaining elements of $\underline{\beta}$ are ℓn p_i , where p_i are the gene frequencies, the estimates of p_i are then proportional to $\exp(\hat{\theta}_i)$, where $\hat{\theta}_i$ are the estimates produced by GLIM. Since $\sum_i \hat{p}_i = 1$, it follows that

$$\hat{p}_i = e^{\hat{\theta}_i} \Big/ \sum_j e^{\hat{\theta}_j} \qquad . \qquad (1)$$

In their treatment of the ABO problem, Thompson and Baker appear to have calculated $\hat{p}$, $\hat{q}$ and $\hat{r}$ from the fitted values $\hat{\mu}_i$ rather than from the $\hat{\theta}_i$. An advantage of using (1) is that it can be used to estimate the (approximate) covariance matrix of the gene frequency estimates from that of the $\hat{\theta}_i$. The covariances of the $\hat{\theta}_i$ are obtained by using $EXTRACT %VC. In general, let $\underline{\hat{\theta}} = (\hat{\theta}_1, \ldots, \hat{\theta}_k)^T$ be a vector of estimates with covariance matrix V, and let $p_i = f_i(\underline{\hat{\theta}})$. By considering Taylor expansions of $f_i(\underline{\hat{\theta}})$ about $\underline{\hat{\theta}}$ it can be seen that the covariance matrix W of $\underline{\hat{p}}$ is given approximately by $W \approx D^T V D$, where D is the k x k matrix with entries

$$d_{ij} = \left(\frac{\partial f_i}{\partial \theta_j} \right)_{\underline{\theta} = \underline{\hat{\theta}}} .$$

The functions f_i will be ratios of sums of exponentials in the examples to be considered here, and with this choice of f_i the above approximation seems to work reasonably well for moderate sample sizes and positive frequencies. A GLIM-3 macro for calculating W is described in Burn and Thomson (1981).

3. ABO AND MNS BLOOD GROUP SYSTEMS

A GLIM-3 listing which fits a loglinear model with composite link function to ABO data is presented in Thompson and Baker (1981). Based on the data $n_A = 179$, $n_B = 35$, $n_{AB} = 6$ and $n_O = 202$, quoted by Kempthorne (1969), we obtain estimates $\hat{p} = 0.2516$, $\hat{q} = 0.0500$, $\hat{r} = 0.6984$ with covariance matrix

$$10^{-4} \times \begin{bmatrix} 2.602 & -0.1484 & -0.2454 \\ -0.1484 & 0.5775 & -0.4291 \\ -0.2454 & -0.4291 & 2.883 \end{bmatrix} \quad .$$

These results are in very good agreement with those of Kempthorne. The deviance is 3.173 on one degree of freedom.

For another example of this kind, consider the MNS blood group system in which there are effectively four alleles denoted Ms, MS, Ns and NS with gene frequencies p_1, p_2, p_3 and p_4, respectively. The manner in which the genotypes combine to produce the phenotypes is described by Cepellini, Siniscalco and Smith (1955) and is summarised in Table 3.1. The 4 x 4 table of expected genotype frequencies in a sample of size n from a random mating population becomes the 16 x 1 vector $\underline{y} = (np_1^2, np_1 p_2, np_1 p_3, \ldots, np_4^2)$. The vector $\underline{\mu}$ of expected phenotype frequencies, of length 6, is related to $\underline{y}$ by $\underline{\mu} = C\underline{y}$, where the 6 x 16 matrix C is deduced from Table 3.1. The loglinear model is then given by $\underline{y} = \exp(\underline{n})$ where $\underline{n} = X\underline{\beta}$, $\underline{\beta} = (\ln n, \ln p_1, \ln p_2, \ln p_3, \ln p_4)^T$ and X is obtained in the same way as in the ABO problem.

TABLE 3.1

	Phenotype	Genotypes
1	MS	MS/MS; MS/Ms
2	Ms	Ms/Ms
3	MNS	MS/NS; MS/Ns; Ms/NS
4	MNs	Ms/Ns
5	NS	NS/NS; NS/Ns
6	Ns	Ns/Ns

The following data are taken from Cepellini et al (1955):

Phenotype:	MS	Ms	MNS	MNs	NS	Ns	Total
Observed frequency:	44	35	62	47	21	21	230

The parameter estimates obtained for these data were $\hat{p}_1 = 0.3800$, $\hat{p}_2 = 0.2004$, $\hat{p}_3 = 0.2892$ and $\hat{p}_4 = 0.1304$, and the estimated covariance matrix is

$$10^{-4} \times \begin{bmatrix} 7.257 & -3.796 & -3.852 & 0.3921 \\ -3.796 & 5.626 & 0.2033 & -2.032 \\ -3.852 & 0.2033 & 6.444 & -2.795 \\ 0.3921 & -2.032 & -2.795 & 4.436 \end{bmatrix}$$

results which agree well with those of Cepellini et al who used the gene counting method. The deviance was 0.5164 on 2 degrees of freedom.

It is clear from the ABO and MNS examples that any multiple allele, single locus problem may be treated in a similar fashion.

4. ESTIMATING THE INBREEDING COEFFICIENT

Some data on the haptoglobin system, obtained from a population in northeastern Brazil, are quoted by Yasuda (1968). In this system there are six phenotypes determined by a single locus with three alleles, which we label G_1, G_2 and G_3. None of the alleles is dominant with respect to the others, so the six phenotypes correspond to genotypes G_1G_1, G_1G_2, G_1G_3, G_2G_2, G_2G_3, G_3G_3, respectively. If p_i denotes the gene frequency of G_i, then the expected phenotype frequencies in a sample of N individuals are Np_1^2, $2Np_1p_2$, $2Np_1p_3$, Np_2^2, $2Np_2p_3$, Np_3^2. The X and C matrices for the random mating model are easily constructed.

The data quoted by Yasuda are as follows.

Phenotype:	1	2	3	4	5	6	Total
Observed frequency:	108	196	429	143	513	559	1948

The estimated gene frequencies are $\hat{p}_1 = 0.2159$, $\hat{p}_2 = 0.2554$ and $\hat{p}_3 = 0.5287$ with estimated covariance matrix

$$10^{-3} \times \begin{bmatrix} 4.344 & -1.415 & -2.929 \\ -1.415 & 4.881 & -3.466 \\ -2.929 & -3.466 & 6.395 \end{bmatrix} .$$

Yasuda's maximum likelihood estimates, obtained by quite a different method, are the same.

The deviance for the random mating model is 7.961 on 3 degrees of freedom, which is significant at the 5% level. This casts some doubt on the random mating hypothesis, so following Yasuda, we consider a model which includes an inbreeding effect.

The meaning of the inbreeding coefficient is discussed by Kempthorne (1969), but for our purposes it is sufficient to note that it is a probability, α (the probability that a pair of like genes are "identical by descent") which modifies the expected phenotype frequencies ($\underline{\mu}$) as shown in Table 4.1. The probability that a pair of genes do not come from a common ancestor is $\beta = 1 - \alpha$.

TABLE 4.1

Phenotype	Expected frequency
1	$N(\alpha p_1 + \beta p_1^2)$
2	$2N\beta p_1 p_2$
3	$2N\beta p_1 p_3$
4	$N(\alpha p_2 + \beta p_2^2)$
5	$2N\beta p_2 p_3$
6	$N(\alpha p_3 + \beta p_3^2)$
Total:	N

Our underlying unobservable frequency table can be regarded as a 3 x 3 x 2 contingency table (Table 4.2), the third factor accounting for the inbreeding.

TABLE 4.2

No common ancestor

	G_1	G_2	G_3
G_1	$N\beta p_1^2$	$N\beta p_1 p_2$	$N\beta p_1 p_3$
G_2	$N\beta p_1 p_2$	$N\beta p_2^2$	$N\beta p_2 p_3$
G_3	$N\beta p_1 p_3$	$N\beta p_2 p_3$	$N\beta p_3^2$

Identical by descent

	G_1	G_2	G_3
G_1	$N\alpha p_1$	0	0
G_2	0	$N\alpha p_2$	0
G_3	0	0	$N\alpha p_3$

Taking for $\underline{y}$ the 12 x 1 vector of non-zero entries in this table
$\underline{y} = (N\beta p_1^2 , N\beta p_1 p_2 , \ldots , N\beta p_3^2 , N\alpha p_1 , N\alpha p_2 , N\alpha p_3)^T$ and
$\underline{\beta} = (\ell n\ N , \ell n\ p_1 , \ell n\ p_2 , \ell n\ p_3 , \ell n\ \alpha , \ell n\ \beta)^T$, we have $\underline{y} = \exp(X\underline{\beta})$ and $\underline{\mu} = C\underline{y}$,
where X (12 x 6) and C (6 x 12) are derived from Tables 4.1 and 4.2.

In this case there are two dependency relations among the parameters, $p_1 + p_2 + p_3 = 1$
and $\alpha + \beta = 1$, so for reasons discussed in Section 2, we fit a loglinear model with
terms corresponding to p_1 , p_2 , p_3 and α , omitting the constant term,%GM. If $\theta_1 , \theta_2 ,$
θ_3 and ϕ denote the corresponding parameters whose estimates are given by the
$DISPLAY E directive, then we have

$$\beta p_i \ \propto\ e^{\theta_i} \quad , \quad i = 1,2,3$$

and $\qquad\qquad \alpha \ \propto\ e^{\phi}$, with the same constant of proportionality.
Hence

$$\hat{p}_i \ =\ e^{\hat{\theta}_i} \Big/ \sum_j e^{\hat{\theta}_j} \quad , \quad i = 1,2,3$$

and $\qquad\qquad \hat{\alpha} \ =\ e^{\hat{\phi}} \Big/ (\sum_j e^{\hat{\theta}_j} + e^{\hat{\phi}})$.

The results obtained from fitting the inbreeding model are as follows:

$$\hat{p}_1 = 0.2157 , \quad \hat{p}_2 = 0.2554 , \quad \hat{p}_3 = 0.5289 , \quad \hat{\alpha} = 0.0431 .$$

The estimated covariance matrix of $[\hat{p}_1, \hat{p}_2, \hat{p}_3, \hat{\alpha}]$ is

$$10^{-4} \times \begin{bmatrix} 0.4485 & -0.1515 & -0.3145 & 0.0175 \\ -0.1515 & 0.5053 & -0.3680 & 0.0123 \\ -0.3145 & -0.3680 & 0.6825 & -0.0298 \\ 0.0175 & 0.0123 & -0.0298 & 2.588 \end{bmatrix} ,$$

and the deviance is 0.819 on 2 degrees of freedom, indicating a more satisfactory fit than the random mating model. Our estimate of α, $\hat{\alpha} = 0.0431$ with standard error 0.0161, is in close agreement with that of Yasuda, which was $\hat{\alpha} = 0.0435$ with S.E. = 0.0160. Note that the gene frequency estimates are virtually unchanged by the inclusion of the inbreeding term (the estimated correlations between $\hat{\alpha}$ and $\hat{p}_i$ are quite small). This orthogonality between the gene frequency parameters and the inbreeding coefficient was also noted by Yasuda.

Other methods of estimating α appear to present difficulties. Yasuda's maximum likelihood scoring method, for instance, can produce negative estimates. An adaptation of the gene counting technique of Cepellini et al (1955) due to Thomson (1982), who has shown the method to be equivalent ot the EM algorithm of Dempster et al (1971), may take a very large number of iterations. The GLIM method, although performing more computation at each step, took just five iterations for the data of this example. The estimate of α is also guaranteed to be positive.

5. CRYPTIC FATHERS

A situation somewhat similar to dominance arises when a genotype is observable in the juvenile of the species but not in the adult. Such is the case at a particular genetic locus in Drosophila pseudoobscura. There are two alleles, which we label A and a, and in adults the genotypes AA and Aa are indistinguishable, while all three genotypes, AA, Aa and aa, are observable in the larvae. It is of interest to evolutionary biologists to estimate the gene frequency of A in wild populations, based on a sample of adults. An experimental procedure that is used in these circumstances is described by Arnold (1981). A random sample of males is mated in the laboratory with females of known aa genotype and the genotypes of the offspring are noted. The genotype is said to be cryptic in the fathers.

Suppose that for each cryptic male in a sample of N, a fixed number, n, of offspring are observed, and let X denote the number of the allele A in the n offspring. Then X can take values $0, 1, \ldots, n$, and the family data on which the inference is to be based consist of observations $x_1, x_2, \ldots, x_N$ of X. The conditional distribution of X, given the genotype of the father, is deduced from the usual Mendelian laws of inheritance:

$$p(x|aa) = 1 \qquad\qquad (x = 0)$$
$$p(x|Aa) = \binom{n}{x} \frac{1}{2^n} \qquad (x = 0,1,\ldots,n) \qquad\qquad (1)$$
$$p(x|AA) = 1 \qquad\qquad (x = n)$$

Let θ denote the gene frequency of A in the wild population and let $\theta' = 1 - \theta$. The distribution of X is then obtained as the mixture

$$p(x|\theta) = p(x|aa)p(aa|\theta) + p(x|Aa)p(Aa|\theta) + p(x|AA)p(AA|\theta) . \qquad (2)$$

As with any mixture, the question of identifiability arises. Arnold (1981) showed that mixtures of this type are indeed identifiable.

The data will be in the form of frequencies N_x , $x = 0,1,\ldots,n$, of the families with x A-genes, where $\sum_x N_x = N$. We can now express the expected frequencies $\mu_x = E(N_x)$ in terms of θ by means of a loglinear model with composite link. As usual, $\underline{\gamma}$ is the vector of expected frequencies in an unobservable table of genotypes in the population: $\underline{\gamma} = (N\theta'^2 , N\theta\theta' , N\theta'\theta , N\theta^2)^T$. (Note that here, as in all our examples, it is convenient to consider Aa and aA genotypes separately, a distinction not usually made by geneticists.) Using (2), we can write

$$u_x = p(x|aa)\gamma_1 + p(x|Aa)\gamma_2 + p(x|aA)\gamma_3 + p(a|AA)\gamma_4 \quad ,$$

and hence $\quad \underline{\mu} = C\underline{\gamma}$, where, from (1), C is given by

$$C^T = \begin{bmatrix} 1 & 0 & 0 & \cdots & 0 & 0 \\ \frac{1}{2^n} & \binom{n}{1}\frac{1}{2^n} & \binom{n}{2}\frac{1}{2^n} & \cdots & \binom{n}{n-1}\frac{1}{2^n} & \frac{1}{2^n} \\ \frac{1}{2^n} & \binom{n}{1}\frac{1}{2^n} & \binom{n}{2}\frac{1}{2^n} & \cdots & \binom{n}{n-1}\frac{1}{2^n} & \frac{1}{2^n} \\ 0 & 0 & 0 & \cdots & 0 & 1 \end{bmatrix} .$$

To complete the specification of the model, let $\underline{\beta} = (\ell n\, N , \ell n\, \theta , \ell n\, \theta')^T$. Then $\underline{\gamma} = \exp(X\underline{\beta})$, where X (4 x 3) is easily obtained. In the data presented by Arnold, the collection size was N = 10 and the family size n = 2. The observed frequencies were $N_0 = 5$, $N_1 = 2$ and $N_2 = 3$. For these data, fitting our loglinear model produces an estimate of $\hat{\theta} = 0.3973$ with estimated standard error 0.1336. The deviance is 0.0925 on one degree of freedom.

With a family size of only 2, as Arnold shows, the maximum likelihood estimate of can be ..ined quite easily by more direct methods. However, Arnold's analysis is based upon collapsing the frequencies $N_1 , \ldots , N_{n-1}$, and likelihood equations based on the original frequencies $N_0 , N_1 , \ldots , N_n$ would not be so simple for larger n.

The main purpose of Arnold (1981) is to present a simple alternative estimator (the 'Dobzhansky estimator') and to show it to be efficient. The GLIM approach, which adapts easily to more complicated genetic situations (e.g. multiple alleles), could be a convenient alternative, producing maximum likelihood (and hence fully efficient) estimates, their covariance matrix and the deviance.

6. ESTIMATING PROBABILITIES ON TREES

The final example shows how to estimate probabilities of events described by an indirectly observable tree diagram instead of the underlying contingency table of the previous problems. Human trisomy (which causes mongolism) can result from nondisjunction at either the first or the second meiotic division, denoted I and II, and this may occur in the gametes of either parent (M or F). Thomson (1981) describes the situation by the following tree diagram, where $p_1 = P(M)$, $q_1 = P(F)$, $p_2 = P(I|M)$, $q_2 = P(II|M)$, $p_3 = P(I|F)$ and $q_3 = P(II|F)$.

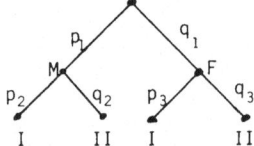

There are thus four types of nondisjunction, and their probabilities will be denoted $\alpha_1, \alpha_2, \alpha_3, \alpha_4$, respectively. Hence

$$\alpha_1 = p_1 p_2 , \qquad \alpha_2 = p_1 q_2 , \qquad \alpha_3 = q_1 p_3 , \qquad \alpha_4 = q_1 q_3 . \qquad (1)$$

Classifying the progeny according to the similarity (or otherwise) of each parent's chromosomes, it turns out that there are 26 progeny types. These are described by Thomson. The observations are the 26 frequencies of occurrence of the progeny types in a sample of N individuals.

Let γ denote the expected frequencies of the four types of nondisjunction, not directly observable: $\gamma = (N\alpha_1, N\alpha_2, N\alpha_3, N\alpha_4)^T$. The probabilities of the 26 progeny types are given by Thomson and they constitute our matrix C:

$$C^T = \begin{bmatrix} 1 & 0 & 0 & 1 & 0 & 1 & 0 & \frac{1}{2} & \frac{1}{2} & 0 & 0 & 0 & 1 & 0 & 1 & 0 & \frac{1}{2} & 0 & \frac{1}{2} & 0 & 0 & 0 & 1 & 0 & 0 & 0 \\ 1 & 0 & \frac{1}{2} & \frac{1}{2} & \frac{1}{2} & 0 & \frac{1}{2} & \frac{1}{2} & \frac{1}{2} & 0 & 1 & 0 & 0 & 0 & 1 & 0 & 0 & \frac{1}{2} & 0 & 0 & \frac{1}{4} & \frac{1}{4} & 0 & 0 & 1 & 0 \\ 0 & 1 & 0 & 1 & \frac{1}{2} & \frac{1}{2} & 0 & 0 & 1 & 0 & 0 & 1 & 0 & 0 & 0 & 1 & 0 & 0 & \frac{1}{2} & 0 & 0 & \frac{1}{2} & 0 & 1 & 0 & 0 \\ 0 & 1 & \frac{1}{2} & \frac{1}{2} & \frac{1}{2} & \frac{1}{2} & 0 & \frac{1}{2} & 0 & \frac{1}{2} & 0 & 1 & 0 & 1 & 0 & 0 & \frac{1}{4} & 0 & 0 & \frac{1}{2} & 0 & \frac{1}{4} & 0 & 0 & 0 & 1 \end{bmatrix} .$$

The expected frequencies of the 26 progeny types are $\mu = C\gamma$. If we take $\beta = (\ln N, \ln \alpha_1, \ln \alpha_2, \ln \alpha_3, \ln \alpha_4)^T$, then we have $\gamma = \exp(\eta)$ where $\eta = X\beta$ and in this case X is the 4 x 5 matrix $[\underline{1}|I]$.

If θ_i , $i = 1,\ldots,4$, denote the parameters estimated by GLIM, then $\hat{\alpha}_i = e^{\hat{\theta}_i}\Big/\sum_j e^{\hat{\theta}_j}$, and from (1) it follows that

$$\hat{p}_1 = (e^{\hat{\theta}_1} + e^{\hat{\theta}_2})\Big/\sum_j e^{\hat{\theta}_j} \quad , \quad \hat{p}_2 = e^{\hat{\theta}_1}\Big/(e^{\hat{\theta}_1} + e^{\hat{\theta}_2}) \quad , \quad \hat{p}_3 = e^{\hat{\theta}_3}\Big/(e^{\hat{\theta}_3} + e^{\hat{\theta}_4})$$

and $\hat{q}_i = 1 - \hat{p}_i$.

Unfortunately, family data for human trisomics is scarce. The data quoted by Thomson (a compilation of data from several studies) are:

Progeny type:	1	2	7	9	10	13	16	19	24
Observed frequency:	2	3	5	4	6	6	13	1	4

The other 17 progeny types had zero frequenices. As noted in Section 2, these zero counts can be expected to lead to difficulties. Even so, we are able to obtain estimates of the p_i:

$$\hat{p}_1 = 0.3233 \quad , \quad \hat{p}_2 = 0.5620 \quad , \quad \hat{p}_3 = 0.7735 \quad ,$$

which agree well with those of Thomson, who used a kind of gene counting technique. However, no doubt because of the zeros in the data, our approximate method for the asymptotic covariance matrix breaks down in this case.

7. DISCUSSION

For all of the applications considered here, there are alternative computational methods. In many cases, the EM algorithm would be a natural choice. All of our models are instances of the loglinear models considered by Haberman (1974), which in turn constitute a subclass of the more general class of "product models" of Haberman (1977). In the latter paper, Haberman discusses alternative methods of solving the likelihood equations. The first is a solution by functional iteration, which amounts to a generalisation of the gene counting technique (a special case of the EM algorithm) developed by Cepellini et al (1955) and Haberman shows that although numerically stable, functional iteration may be very slow to converge. Of the other methods considered by Haberman, he favours a Newton-Raphson/scoring method, which is essentially what GLIM uses.

ACKNOWLEDGEMENT

The author is grateful to R Thomson of Brighton Polytechnic for suggesting some of the examples and for many useful comments.

REFERENCES

ARNOLD, J. (1981). Statistics of natural populations. I: Estimating an allele probability in cryptic fathers with a fixed number of offspring. *Biometrics, 37,* 495-504.

BISHOP, Y.M.M., FIENBERG, S.E. and HOLLAND, P.W. (1975). *Discrete Multivariate Analysis.* M.I.T. Press.

BURN, R. and THOMSON, R. (1981). A macro for calculating the covariance matrix of functions of parameter estimates. *The GLIM Newsletter, No. 5.* Oxford: Numerical Algorithms Group.

CEPELLINI, R., SINISCALCO, M. and SMITH, C.A.B. (1955). The estimation of gene frequencies in a random mating population. *Ann. Hum. Genet., 20,* 97-115.

DEMPSTER, A.P., LAIRD, N.M. and RUBIN, D.B. (1977). Maximum likelihood from incomplete data via the EM algorithm. *J.R. Statist. Soc. B.,* 39, 1-38.

ELANDT-JOHNSON, R.C. (1971). *Probability Models and Statistical Methods in Genetics.* New York: Wiley.

HABERMAN, S.J. (1974). Loglinear models for frequency tables derived by indirect observation: maximum likelihood equations. *Ann. Statist., 2,* 911-924.

HABERMAN, S.J. (1977). Product models for frequency tables involving indirect observation. *Ann. Statist., 5,* 1124-1147.

KEMPTHORNE, O. (1969). *An Introduction to Genetic Statistics.* Ames, Iowa : Iowa State University Press.

NELDER, J.A. and WEDDERBURN, R.W.M. (1972) Generalised linear models. *J.R. Statist. Soc. A, 135,* 370-383.

THOMSON, R. and BAKER, R.J. (1981). Composite link functions in generalised linear models. *Appl. Statist., 30,* 125-131.

THOMSON, R. (1981). Estimating the origins of human trisomics and triploids. *Ann. Hum. Genet., 45,* 65-78.

THOMSON, R. (1982). PhD Thesis (to appear). *University College, London.*

YASUDA, N. (1968). Estimation of the inbreeding coefficient from phenotype frequencies by a method of maximum likelihood scoring. *Biometrics, 24,* 915-934.

USE OF THE COMPLEMENTARY LOG-LOG FUNCTION TO DESCRIBE DOSE-RESPONSE

RELATIONSHIPS IN INSECTICIDE EVALUATION FIELD TRIALS

KATHLEEN PHELPS

National Vegetable Research Station
Wellesbourne, Warwick CV35 9EF

SUMMARY

Generalised linear modelling techniques are used to provide a standard method for
the analysis of data from insecticide trials. The complementary log-log function
provides very well-fitting dose-response curves (details to be published elsewhere).
The offset facility allows a constant term to be incorporated in the model to adjust
for the base level of insect attack which varies from experiment to experiment.

Each insecticide dose was replicated three times. By considering dose both as a
variable and a factor the residual deviance can be partitioned into a "lack of fit"
and a "pure error" term thus overcoming the problems of interpreting the absolute
residual deviance which is related to the parameter value. The partitioned deviances
can then be used to assess the goodness of fit of the line.

Keywords: Dose-response; complementary log-log; granular insecticide formulations;
partitioning of residual deviances.

1. INTRODUCTION

Field trials are done at the National Vegetable Research Station to evaluate the
performance of granular formulations of insecticides against carrot fly on carrots
(Thompson and Wheatley, 1977). Important features of these trials are that the insecticides are applied to plots in continuous logarithmically increasing doses and
that a large number of untreated check plots, arranged in a systematic grid, is included to measure accurately the base level of insect infestation over the area of
the experiments. Plots are divided into 10 equal-length subplots at harvest for the
assessment of carrot fly damage on the carrot roots. These trials differ from conventional laboratory toxicological experiments in that insect-mortality following
insecticide treatments cannot be measured directly. Instead, an indirect estimate
of the percentage of individuals surviving treatments is obtained by comparing the
level of insect damage on the treated and check subplots. The complementary log-log
function (Mather, 1949) provides a very convenient linear model for describing the
data (Phelps, 1979), the relevant equation for a particular insecticide being:-

$$\ln(-\ln q_{2i}) = \ln(-\ln q_1) + \alpha + \beta \log(d_i) \dots\dots\dots\dots\dots\dots(1)$$

where

d_i is the mean dose of insecticide applied to subplot i

q_{2i} is the proportion of undamaged roots on a subplot treated with dose d_i

q_1 is the proportion of undamaged roots obtained from the check subplots in the experiment

alpha and beta are parameters.

The comparison between treated and check subplots is an important feature of this model. Inclusion of the term $\ln(-\ln q_1)$ makes alpha independent of the base level of infestation in a particular experiment thus facilitating comparisons between experiments. Furthermore, it allows the results to be expressed subsequently in terms of relative population densities of carrot fly on treated and check subplots (Wheatley and Freeman, 1982).

The model has been found to fit data on the biological performance of many insecticides very well (details to be published elsewhere). Results from a trial to assess the performance of insecticides against carrot fly on carrots are used in this paper to illustrate the appropriate GLIM program. Some of the problems involved in assessing how well the model fits the data for some of the individual treatments are discussed.

2. METHODS

2.1 THE DATA USED IN THE MODEL

The data were obtained in a carrot trial done at the National Vegetable Research Station. The trial consisted of 3 replicate blocks, each containing 12 plots treated with different insecticides and a systematic grid of 7 untreated check plots. The numbers of damaged and undamaged carrots in each subplot were recorded. The first and last subplots of each plot were excluded from the statistical analysis because of "end effects" in the application of insecticide and carrot fly attack.

Thus the data used in fitting the model for each insecticide were based on 3 replicates of 8 doses. Data from the check subplots were used to calculate a mean value of proportion undamaged carrots for each replicate.

2.2 SPECIFICATION OF THE MODEL IN GLIM TERMINOLOGY

Equation (1) was fitted very easily in GLIM with a binomial error structure, r being the number of damaged carrots and n the total number of carrots per subplot. The logarithm of the mean dose of insecticide applied to a subplot was the x-variable. The term $\ln(-\ln q_1)$ representing the base level of infestation was included by using the offset facility.

Although the obvious data to use in fitting the model were the total numbers of carrots from the 3 replicates of each treatment dose the deviances so obtained were difficult to interpret. However, the replication at each insecticide dose allowed the residual deviance (22df) to be separated into two components, one measuring the adequacy of the line in describing the dose response curve (6df), the other measuring the magnitude of the random variation between replicates (16df).

In order to partition the deviance two models were fitted, both treating the data as three replicate samples at each dose. The between-replicate residual deviance was obtained in GLIM by fitting "dose" as a qualitative factor with 8 levels. The lack-of-fit residual deviance was obtained by subtracting the between-replicate residual deviance from the total residual deviance after fitting the line. To make the partitioning meaningful the offset variable had to take different values for each replicate in both the models, otherwise any differences in insect attack between the 3 replicates would have been included in the between-replicate residual deviance.

3. RESULTS AND DISCUSSION

3.1 PRESENTATION OF RESULTS

The data from field trials such as that described here are processed routinely by an integrated system involving an M50F hand-held data logger, GENSTAT and a Benson plotter. The GLIM program illustrates the model-fitting part of the procedure, normally performed in GENSTAT, for one of the insecticides (phorate). Although graphs are usually drawn on the transformed scale with the ordinates adjusted according to the mean value of the offset variable so that results from different field trials can be compared directly they are shown here (Fig. 1) on the unadjusted scale for clarity. The final part of the GLIM output shows the values plotted on the graph.

Results are tabulated for only 4 of the 12 insecticides used in the trial (Tables 1 and 2). The other 8 were omitted because they were not sufficiently effective against carrot fly to be of further interest. The actual doses applied depended on the individual insecticide formulations but, for the present purposes, they are labelled 1 (low) to 8 (high). The offset variable, $\ln(-\ln q_1)$ held the values 0.97, 0.86 1.03 for replicates 1 to 3 respectively.

Although significance testing is not very important in this type of work some assessment of the fit of the model was required. Three possible methods are illustrated here and some of their advantages and disadvantages are discussed.

3.1.1 GRAPHICAL METHOD

The graphs in Fig. 1 provided an excellent visual comparison of insecticide performance. Thus it was immediately obvious that the dose-response line for carbofuran was much steeper than those for the other insecticides and that at the doses applied carbofuran and phorate were more effective in controlling carrot fly damage than were the other two insecticides. The graphs would also indicate any systematic non-linear trends in the data.

One problem with using the complementary log-log function is that the transformed scale is very difficult to assimilate. For instance, after transformation, the diff-

```
$C...PROGRAM TO ILLUSTRATE HOW MODEL IS FITTED
$C...INSECTICIDE:-PHORATE
$UNITS 24
$DAT LDOS OFF DAM TOT $READ
1.52  0.97  10  35
1.64  0.97  16  42
1.76  0.97   8  50
1.88  0.97   6  42
2.00  0.97   9  35
2.12  0.97   9  42
2.24  0.97   1  32
2.36  0.97   2  28
1.52  0.86  17  38
1.64  0.86  10  40
1.76  0.86   8  33
1.88  0.86   8  39
2.00  0.86   5  47
2.12  0.86  17  42
2.24  0.86   6  35
2.36  0.86   4  35
1.52  1.03  10  34
1.64  1.03  10  38
1.76  1.03   5  36
1.88  1.03   3  35
2.00  1.03   2  49
2.12  1.03   1  40
2.24  1.03   3  22
2.36  1.03   2  31
$C...
$C...OFFSET VARIABLE REPRESENTS BASE LEVEL OF INFESTATION
$C...(NOTE SEPARATE VALUE FOR EACH REP)
$OFFSET OFF
$ALIAS
$C...
$C...NULL MODEL AND REGRESSION LINE
$YVAR DAM $ERR B TOT $LINK C
$FIT
$FIT LDOS
            SCALED
   CYCLE   DEVIANCE      DF
     4      90.53        23

$DISP ER $CAL LINP=%LP
            SCALED
   CYCLE   DEVIANCE      DF
     4      61.79        22

        ESTIMATE      S.E.        PARAMETER
    1    0.5170       0.5651      %GM
    2   -1.606        0.3060      LDOS
    SCALE PARAMETER TAKEN AS      1.000
```

UNIT	OBSERVED	OUT OF	FITTED	RESIDUAL
1	10	35	11.19	-0.4305
2	16	42	11.43	1.585
3	8	50	11.52	-1.183
4	6	42	8.160	-0.8425
5	9	35	5.712	1.504
6	9	42	5.739	1.465
7	1	32	3.652	-1.474
8	2	28	2.663	-0.4271
9	17	38	11.09	2.110
10	10	40	9.906	0.3452E-01
11	8	33	6.902	0.4697
12	8	39	6.862	0.4784
13	5	47	6.934	-0.7953
14	17	42	5.180	5.547
15	6	35	3.601	1.335
16	4	35	2.997	0.6057
17	10	34	11.41	-0.5130
18	10	38	10.88	-0.3154
19	5	36	8.741	-1.454
20	3	35	7.175	-1.748
21	2	49	8.446	-2.438
22	1	40	5.778	-2.149
23	3	22	2.656	0.2249
24	2	31	3.121	-0.6692

```
$C...MODEL WITH DOSE AS QUALITATIVE FACTOR
$C...USED BOTH TO CALCULATE BETWEEN REP RESIDUAL
$C...AND TO CALCULATE WEIGHTED MEANS AT EACH DOSE
$FACT FDOS 8
$CAL FDOS=%GL(8,1)
$FIT FDOS
$EXTRACT %PE
        SCALED
 CYCLE  DEVIANCE       DF
   4     49.29         16

$DISP E $
       ESTIMATE     S.E.       PARAMETER
  1    -3.372      0.3539      %GM
  2     1.559      0.3908      FDOS(1)
  3     1.387      0.3915      FDOS(2)
  4     0.7716     0.4160      FDOS(3)
  5     0.5745     0.4293      FDOS(4)
  6     0.3774     0.4334      FDOS(5)
  7     1.007      0.4033      FDOS(6)
  8     0.2998     0.4748      FDOS(7)
  9      ZERO      ALIASED     FDOS(8)
  SCALE PARAMETER TAKEN AS     1.000

$CAL COEF=%PE
$VARI 8 PLIN LD CF
$DAT R1 $READ
1 2 3 4 5 6 7 8 0 0 0 0 0 0 0 0 0 0 0 0 0 0 0  0 0
$DAT 9 C1 $READ
0 2 3 4 5 6 7 8 9
$C...
$C:..PLIN HOLDS LINEAR PREDICTOR VALUES FOR REPLICATE 1
$CAL PLIN(R1)=LINP
     :COEF(C1)=COEF(C1)+COEF(1)
$DAT  9 C $READ
0 1 2 3 4 5 6 7 8
$CAL CF(C)=COEF
$C...
$C:..ADD MEAN OFFSET VALUE TO WEIGHTED MEANS
$C...ADJUST LINEAR PREDICTOR TO MEAN OFFSET VALUE
$CAL CF=CF+0.95
$CAL PLIN=PLIN-0.97+0.95
$C...
$C:..CF HOLDS OBSERVED POINTS FOR PLOTTING
$C:..PLIN HOLDS FITTED LINE
$LOOK CF PLIN
$STOP
  1   -0.8631      -0.9741
  2   -1.035       -1.167
  3   -1.650       -1.360
  4   -1.848       -1.552
  5   -2.045       -1.745
  6   -1.415       -1.938
  7   -2.122       -2.130
  8   -2.422       -2.323
```

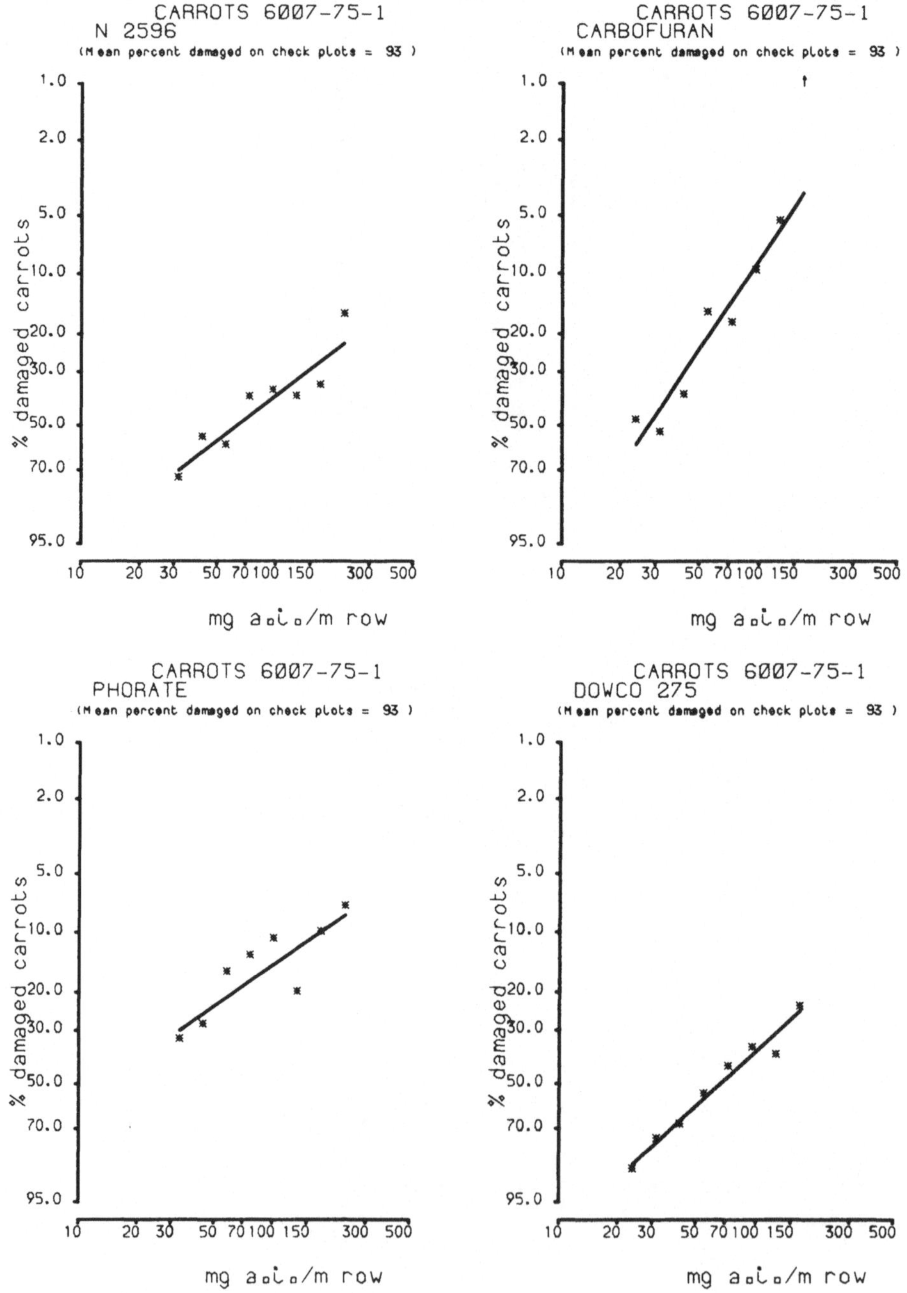
CARROTS 6ØØ7-75-1
N 2596
(Mean percent damaged on check plots = 93)
CARROTS 6ØØ7-75-1
CARBOFURAN
(Mean percent damaged on check plots = 93)
CARROTS 6ØØ7-75-1
PHORATE
(Mean percent damaged on check plots = 93)
CARROTS 6ØØ7-75-1
DOWCO 275
(Mean percent damaged on check plots = 93)
% damaged carrots
1.0
2.0
5.0
10.0
20.0
30.0
50.0
70.0
95.0
10
20
30
50
70
100
150
300
500
mg a.i./m row

Fig. 1 Plots of $-\ln(-\ln q_{2i})$ against $\log_{10} d_i$

(↑ and ↓ represent points falling beyond the axes)

erence between 40 and 50 percent damaged carrots is approximately the same as that
between 0.3 and 0.4 percent. In addition, the weighting given to the points during
the fitting procedure is difficult to show on a graph. Consequently, graphical presentation can be very misleading. Fig. 1 illustrates some of the problems; upward and
downward arrows indicating points beyond the scale. Inspection of these graphs suggests that the lines fit very well with the exception of one very "wild" point on the
carbofuran graph and apparently rather variable results with phorate. However, these
apparently poor-fitting points appear poor only because of the log-log scale used.
In fact the wild point on the carbofuran graph represents an observed value of 1 damaged carrot out of a total of 144 compared with a predicted number of 6.6. Similarly
the phorate results appear more variable than those of Dowco 275 and N2596 because
they lie on a more "sensitive" part of the vertical scale.

3.1.2 PARAMETER ESTIMATION AND ANALYSIS OF DEVIANCE

Values of the parameters alpha and beta (Table 1) give statistical confirmation
of the graphical comparisons of insecticide performance. Interpretation of these two
parameters is beyond the scope of this paper.

Table 1. Regression coefficients and analysis of deviance

Insecticide	α	β	Overall (22 df)	Lack of fit (6 df)	Residual (16 df)
Phorate	0.51	−1.60	2.82	2.08	3.10
N2596	1.88	−1.76	3.52	2.81	3.78
Dowco 275	1.74	−2.13	3.94	0.81	5.11
Carbofuran	3.74	−3.49	4.70	4.02	4.95

The overall residual deviances for the four models fitted (Table 1) gave little
indication of the relative goodness-of-fit of the models because they depended on the
value of the parameters of the model (Wood, 1981). Thus low values of residual deviance would be expected with low levels of insect damage and vice versa. A better
indication of the goodness-of-fit of the models was obtained from the partitioned
deviances (Table 1). The mean deviances from the original regression lines, and their
two components were typical of those obtained in other experiments. Between-replicate
deviances (16 df) were generally correlated with percentage damaged carrots, as expected, unlike lack-of-fit deviances (6 df) which varied haphazardly. The results for
the insecticide Dowco 275 illustrate this, the high value of the between-replicate
deviance being due partly to the high values of percentage damaged carrots. Thus the
between-replicate residual deviance, like the overall residual deviance, provides
little information about the fit of the model. However, comparison of the two deviances yielded more information. If the between-replicate residual deviance were
smaller than the overall residual deviance it would imply that further parameters were

required for the model to describe the data adequately. In fact the between-replicate mean deviance was always larger. Thus there was no evidence of lack-of-fit of
the regression line. This approach differs from the usual one for normal errors where
the lack-of-fit term would be compared with the between-replicate term (Draper and
Smith, 1966). F-ratios comparing the lack-of-fit term with the between-replicate
term would be meaningless here because they would be highly influenced by the expected fluctuation in the residual deviance due to the variation in parameter values.

Use of the deviances to measure goodness of fit suffers from the same problems as
graphical methods. Lack of linearity can be shown easily but the variability is assessed with more difficulty.

3.1.3 COMPARISONS ON THE UNTRANSFORMED SCALE

The observed percentages of damaged carrots can be compared with those predicted
from the fitted line (Table 2). This approach has an intuitive appeal in that the
percentage scale can be comprehended readily. It can be used in conjunction with the
other methods to judge the practical importance of deviations from the fitted line.

Table 2.   Observed and predicted percentages of damaged carrots (totalled
           over 3 replicates)

Dose level per sub-plot	Phorate		N2596		Dowco 275		Carbofuran	
	Obs.	Pred.	Obs.	Pred.	Obs.	Pred.	Obs.	Pred.
1 (low)	34.6	31.5	73.8	70.0	86.2	84.0	48.9	58.8
2	30.0	26.8	55.9	61.9	74.1	75.5	54.0	44.1
3	17.6	22.8	59.6	54.4	68.1	66.1	39.8	32.2
4	14.6	19.1	41.2	47.4	54.6	56.4	17.2	22.2
5	12.4	16.4	37.7	40.2	44.6	48.0	19.0	15.1
6	21.8	13.5	40.0	34.1	45.3	39.7	11.0	10.3
7	11.2	11.1	36.4	28.8	40.0	32.4	6.3	6.9
8 (high)	8.5	9.3	18.1	24.4	25.0	26.0	0.7	4.6

CONCLUSIONS

The introduction of GLM techniques enabled data which had previously proved difficult
to handle satisfactorily to be analysed statistically. The methods described in this
paper are now used routinely.

Although the complementary log-log function was considered originally because of the
convenient form of Equation 1, it soon became obvious that it was an ideal transformation for these data. A feature of the results from insecticide evaluation trials is
that only insecticides which achieve a high level of protection are suitable for further development. It is sensible, therefore, to express the results of experiments
on a scale which emphasizes low percentages of damaged carrots. Differences in high

percentages are of considerably less interest. In fact insecticides which do not achieve an adequate level of control are omitted from the statistical analysis.

ACKNOWLEDGEMENTS

I thank Dr A.R. Thompson for the use of his data and for helpful discussions.

REFERENCES

DRAPER, N. and SMITH, H. (1966). Applied Regression Analysis. New York: Wiley, pp 62-64.
MATHER, K. (1949). The analysis of extinction time data in bioassay. Biometrics 5, 127-143.
PHELPS, K. (1979). Statistics: Proposed method for analysis of insecticide trials. Report of the National Vegetable Research Station for 1978, p. 122.
THOMPSON, A.R. and WHEATLEY, G.A. (1977). Design of trials for evaluating insecticide against some pests of vegetables. Pesticide Science 8, 418-427.
WHEATLEY, G.A. and FREEMAN, G.H. (1982). A method of using the proportions of undamaged carrots or parsnips to estimate the relative population densities of carrot fly (Psila rosae) larvae, and its practical application. Annals of Applied Biology 100, 229-244.
WOOD, J. (1981). Letter to GLIM Newsletter 5.

GLIM FOR PREFERENCE

C.D. Sinclair
Department of Statistics
St. Andrews University
Fife Scotland

SUMMARY

Many types of multiple comparisons data can be described by multiplicative models, consequently GLIM can be used to fit log-linear models to such data, and to produce not only estimates of the parameters of interest, but also standard errors and tests of significance. Moreover as the data consists of counts the adequacy of the model can be tested in an objective way. Three examples from the literature are analysed, fitting the basic Bradley-Terry model and a modified Bradley-Terry model with ties and order effects. The parametrisation used in the case of ties is claimed to be more convenient than that of Fienberg. A test of linearity is also included.

Keywords : *Bradley-Terry Model, Paired Comparisons, Preference Testing, Log-Linear Model.*

1. THE BRADLEY-TERRY MODEL

Bradley and Terry (1952), and Zermelo (1929) proposed a model to describe the following experiment.

The t items T_i, $i=1, \ldots, t$ are compared in pairs, by n_{ij} judges who each express preferences between T_i and T_j, $N = \sum_{i<j} \sum n_{ij}$.

Independence is assumed for ratings of the same pair by different judges and for ratings of different pairs by the same judge.

The probability that item T_i is preferred to item T_j is

$$P_{i/i,j} = \frac{\Pi_i}{\Pi_i + \Pi_j} \qquad i \neq j = 1, \ldots, t$$

Π_i is the worth of item T_i, $i=1, \ldots, t$.

$\Pi_i > 0$ $i=1, \ldots, t$. $[P_{i/i,j} \geq 0]$.

$\sum_{i=1}^{t} \Pi_i = 1$ gives an arbitrary scaling.

If X_{ij} is the number of times that T_i is preferred to T_j in the n_{ij} comparisons of this pair

$$X_{ij} \cap \text{Binomial } (n_{ij}, P_{i/i,j})$$

Thus for maximum likelihood estimation of the parameters Π_i we wish to

Maximise
w.r.t. $\Pi_1, \ldots, \Pi_t$ $\displaystyle\prod_{\substack{i<j \\ =1}}^{t} \binom{n_{ij}}{x_{ij}} \left(\frac{\Pi_i}{\Pi_i + \Pi_j}\right)^{x_{ij}} \left(\frac{\Pi_j}{\Pi_i + \Pi_j}\right)^{n_{ij}-x_{ij}}$

GLIM can be used to carry out an iterative maximum likelihood fit, provided that the binomial parameter can be formulated in such a way that its logit can be written as a linear model.

Now logit $\left(\dfrac{\Pi_i}{\Pi_i + \Pi_j}\right) = \ell n \ (\Pi_i/\Pi_j) = \lambda_i - \lambda_j$

where $\lambda_j = \ell n \ \Pi_j$ and the linear predictor is $n_{ij} = \lambda_i - \lambda_j$.

Example 1.

In Atkinson's (1972) paper on testing the linear logistic and Bradley-Terry models, he quotes an example of Bradley's on investigating the effect of monosodium glutamate on the flavour of apple sauce. Treatments 1,2,3 are increasing amounts of additive, treatment 4 is a control with no additive. Four independent comparisons were made of each of the six pairs.

TABLE 1

Data and model for effect of monosodium glutamate on the flavour of apple sauce

Items Compared	No. of times i preferred to j	No. of times i compared to j	linear predictor	columns of design matrix			
i j	x_{ij}	n_{ij}	n_{ij}	$\underline{\ell}_1$	$\underline{\ell}_2$	$\underline{\ell}_3$	$\underline{\ell}_4$
1 2	3	4	$\lambda_1 - \lambda_2$	1	-1		
1 3	3	4	$\lambda_1 - \lambda_3$	1		-1	
1 4	3	4	$\lambda_1 - \lambda_4$	1			-1
2 3	3	4	$\lambda_2 - \lambda_3$		1	-1	
2 4	4	4	$\lambda_2 - \lambda_4$		1		-1
3 4	0	4	$\lambda_3 - \lambda_4$			1	-1

The explanatory variables $\underline{\ell}_1$ are vectors with i,jth element the coefficient of λ_i in the linear predictor. Note the lack of identifiability. The design matrix is not of full rank. Addition of an arbitrary quantity to every λ_i will not affect the values of the linear predictors, since they are differences between λ's.

In the printout below the goodness of fit of the Bradley-Terry model as assessed by the deviance, is 6.937, on three degrees of freedom. As the 0.10 and 0.05 points of the chi-squared distribution are 6.25 and 10.82 respectively, this result casts some doubt on the adequacy of the model. A display of the parameter estimates $\hat{\lambda}_i$, reveals that, apart from an arbitrary constant, they are identical to the $\hat{\rho}_i$ obtained by Atkinson.

```
$C APPLE SAUCE , ATKINSON , BIOMETRIKA 1972
$UNITS 6 $DAT I J X $READ
1 2 3 1 3 3 2 3 3 1 4 3 2 4 4 3 4 0
$CAL N=4 $YVAR X $ERR B N
$CAL L1=%EQ(I,1)-%EQ(J,1)
:L2=%EQ(I,2)-%EQ(J,2)
:L3=%EQ(I,3)-%EQ(J,3)
:L4=%EQ(I,4)-%EQ(J,4)
$FIT L1+L2+L3+L4-%GM
$C  SIGNIFICANT AT 10% AS IT EXCEEDS 6.25 , BUT NOT AT 5%
        SCALED
 CYCLE  DEVIANCE      DF
    4    6.937        3

$DIS E $
        ESTIMATE     S.E.     PARAMETER
    1    1.211      0.8390    L1
    2    0.8943     0.8060    L2
    3   -0.9961     0.8680    L3
    4    ZERO       ALIASED   L4
    SCALE PARAMETER TAKEN AS  1.000
```

```
$C NOW CALCULATE ESTIMATES OF WORTH PARAMETERS
$EXTR %PE $CAL G=%EXP(%PE):T=%CU(G):M=G/T(4)
$C NOW CALCULATE APPROXIMATE STANDARD ERRORS
$FIT L1+L2+L3-%GM $EXTR %VC $CAL VC=%VC
           SCALED
   CYCLE  DEVIANCE      DF
      4     6.937        3

$CAL V1=I:V2=J-1:V12=%EQ(V1,V2)
$MACRO VAR
$CAL CVM=V12*M(%I)**2*(M(V2)**2+%EQ(%I,V2))
:CVM=CVM+(1-V12)*2*M(%I)**2*M(V2)*(M(V1)-%EQ(V1,%I))
$CAL VM=CVM*VC
$CAL SM(%I)=%SQRT(%CU(VM))
$CAL %I=%I+1 : %Q=%LT(%I,3)
$ENDMAC
$CAL %I=1 :%Q=1 $WHILE %Q VAR
$CAL VM=VC*M(%I)**2*M(V1)*M(V2)*(2-V12)
$CAL SM(%I)=%SQRT(%CU(VM))
$LOO 1 %I M SM
$C PARAMETER ESTIMATES AND APPROXIMATE STANDARD ERRORS $
      1   0.4681      0.4191
      2   0.3410      0.3392
      3   0.5149E-01  0.5451E-01
      4   0.1394      0.8451E-01
$C TEST FOR NON-LINEARITY
$CAL SQ=%LP**2:%D=%DV $FIT +SQ $CAL %R=%D-%DV
           SCALED
   CYCLE  DEVIANCE      DF
      5     1.944        2

$PRI %R
$C  REDUCTION IN DEVIANCE FOR ONE DEGREE OF FREEDOM
      4.993
$RETURN
$END
```

TABLE 2

Parameter Estimates and their standard errors

i	$\hat{\rho}_i$	$\hat{\lambda}_i$	S.E.$(\hat{\lambda}_i)$	† arbitrary value
1	3.21	1.21	0.84	
2	2.89	0.89	0.81	$\tilde{\lambda}_i = \hat{\rho}_i - 2$
3	1.00	-1.00	0.87	
4	2†	ZERO	ALIASED	

The standard errors reported along with the parameter estimates refer to the parameters λ_i in the linear model. In order to obtain estimates of the worth parameters $\hat{\Pi}_i$, and their standard errors, we recall that

$$\Pi_i = e^{\lambda_i + c} \qquad i = 1, 2, 3, 4$$

where the arbitrary constant c can be chosen so that $\sum_{i=1}^{4} \Pi_i = 1$

i.e $\quad \hat{\Pi}_i \quad = \quad \dfrac{e^{\hat{\lambda}_i}}{\sum\limits_{i=1}^{t} e^{\hat{\lambda}_i}} \quad = \quad \dfrac{A_i}{D} \quad$ say, $\quad i=1, \ldots, 4$

Hence

$$Var[\hat{\Pi}_i] \quad = \quad Var(A_i) \quad + \quad \dfrac{(EA_i)^2}{(ED)^4} \quad Var(D) \quad - \quad \dfrac{2(EA_i)}{(ED)^3} \quad Cov\ (A_i, D)$$

where

$$Var[A_i] \quad = \quad Var[e^{\hat{\lambda}_i}] \quad = \quad e^{2\hat{\lambda}_i} \quad Var(\hat{\lambda}_i)$$

$$Cov[e^{\hat{\lambda}_i}, e^{\hat{\lambda}_j}] \quad = \quad e^{\hat{\lambda}_i + \hat{\lambda}_j} \quad cov(\hat{\lambda}_i, \hat{\lambda}_j)$$

$$Cov[A_i, D] \quad = \quad \sum_j e^{\hat{\lambda}_i + \hat{\lambda}_j} \quad cov(\hat{\lambda}_i, \hat{\lambda}_j)$$

$$Var[D] \quad = \quad \sum\sum_{ij} e^{\hat{\lambda}_i + \hat{\lambda}_j} \quad cov(\hat{\lambda}_i, \hat{\lambda}_j)$$

After the linear model has been fitted in GLIM, the parameter estimates $\hat{\lambda}_i$ i=1,2,3 are in the system vector %PE, and $Var(\hat{\lambda}_i)$, $Cov(\hat{\lambda}_i, \hat{\lambda}_j)$ are in system vector %VC from which they can be extracted and used in calculations. The results are shown in the printout, following the statement

$$\$LOO \quad 1 \quad \%I \quad M \quad AM.$$

Tests of hypothesis about the equality of any two worth parameters Π_i, Π_j can be carried out either by fitting a model with a reduced number of parameters, and suitably modified design matrix, or by testing the equivalent hypothesis about the linear model parameters λ_i, λ_j, perhaps using the standard errors of differences of parameters as produced by $ DIS S.

Atkinson's primary purpose was to propose a test of the linearity of the logistic model. His test statistic is based on the rate of change of the log-likelihood with repsect to the parameter k in

$$logit\ (P_i) = \beta(x_i - \mu)\ \{1 + e^{\alpha(x_i - \mu)}\}\ \{1 + e^{k(x_i - \mu)^2}\}$$

where x_i is the dose, and μ is the mean dose. We have logit $(P_{i/i,j}) = n_{ij} = \lambda_i - \lambda_j$ so an additive linear model anticipates that an increase in $\lambda_i - \lambda_j$ will produce a linear increase in logit $(P_{i/i,j})$.

One way of introducing non-additivity is to include in the model a term in $\lambda_i \lambda_j$. This is equivalent to the Rojas form of Tukey's one degree of freedom for non-additivity, and can be conveniently done by adding to the model a term in the square of the linear predictor

$$\eta_{ij}^2 = (\lambda_i - \lambda_j)^2 = \lambda_i^2 + \lambda_j^2 - 2\lambda_i\lambda_j.$$

In the example this produces a drop in deviance of 4.99 at the expense of 1 degree of freedom, $\chi^2(1,0,05) = 3.84$, supporting Atkinson's conclusion that the data exhibit a significant departure from linearity.

2. THE INCOMPLETE CONTINGENCY TABLE APPROACH

The preference for i, X_{ij}, and for j, X_{ji} $(=n_{ij}-X_{ij})$, can be regarded as forming a contingency table, from which the diagonal entries are missing. The random variables X_{ij} are assumed to follow the Poisson distribution. Conditioning on a fixed total $m = \sum_i \sum_j X_{ij}$ the X_{ij} have the multinomial distribution. Their expected values μ_{ij} have multiplicative model, hence use of the logarithmic link function results in a linear model for n_{ij}.

The Bradley-Terry model can be expressed as

$$P_{i/i,j} = \frac{\Pi_i}{\Pi_i + \Pi_j} = \frac{(\Pi_i/\Pi_j)^{\frac{1}{2}}}{(\Pi_i/\Pi_j)^{\frac{1}{2}} + (\Pi_j/\Pi_i)^{\frac{1}{2}}}$$

similarly $P_{j/i,j} = \dfrac{(\Pi_j/\Pi_i)^{\frac{1}{2}}}{(\Pi_j/\Pi_i)^{\frac{1}{2}} + (\Pi_i/\Pi_j)^{\frac{1}{2}}}$

so if $m_{i/i,j}$ is the expected number of comparisons in which i is preferred to j

$$m_{i/i,j} = \frac{m(\Pi_i/\Pi_j)^{\frac{1}{2}}}{(\Pi_i/\Pi_j)^{\frac{1}{2}} + (\Pi_j/\Pi_i)^{\frac{1}{2}}}$$

$m_{i/i,j}$ has loglinear model :-

$$\ln m_{i/i,j} = \mu_{ij} + \tfrac{1}{2}\lambda_i - \tfrac{1}{2}\lambda_j \tag{1}$$

$$\ln m_{j/i,j} = \mu_{ij} + \tfrac{1}{2}\lambda_i - \tfrac{1}{2}\lambda_j \tag{2}$$

where $\mu_{ij} = \mu_{ji} = \ln m - \ln [(\Pi_i/\Pi_j)^{\frac{1}{2}} + (\Pi_j/\Pi_i)^{\frac{1}{2}}]$ and $\lambda_i = \ln \Pi_i$

Using GLIM with Poisson error and link L where the design matrix consists of column vectors with suitable entries under μ_{ij}, λ_i etc., these λ_i can be estimated by iterative maximum likelihood.

Since the number of observations is $t(t-1)$, while the number of parameters is $(t-1)(t+2)/2$ and the number of degrees of freedom for error are $(t-1)(t-2)/2$, a minimum of three items, and preferably at least five items are required if any hypothesis testing is intended.

Example 2.

Imrey Johnson & Koch (1976). A sample of 477 White North Carolina women, under 30 years of age, married to their first husbands, were asked which family size they preferred, out of a pair of family sizes chosen from 0,1,2, ..., 6. Each woman was asked about only 1 pair, pairs were randomly assigned to women.

TABLE 3

Desired family size : choice i or j

$_j$i	0	1	2	3	4	5	6
0		17	22	22	15	26	25
1	2		19	13	10	9	11
2	1	0		11	11	6	6
3	3	1	7		6	2	6
4	1	10	12	13		4	0
5	1	11	18	15	17		11
6	2	13	20	22	14	12	

[Out of the 19 women who were asked which of the family sizes 0 and 1 they preferred, 2 women preferred family size 0 and 17 preferred 1.]

In the printout overleaf, the Bradley-Terry model is seen to be inadequate for this data, as it results in a deviance of 33.65 on 15 degrees of freedom.

Imrey Johnson and Koch used a non-iterative least squares technique, with weights inversely proportional to variance, to fit a linear model to

$$u_{ij} = \text{logit} \ (P_{i/i,j}) = \ln \ (P_{i/i,j}/P_{j/i,j})$$

with $E[u_{ij}] = \ln \ (\Pi_i) - \ln \ (\Pi_j)$

and $\text{Var}[u_{ij}] = \dfrac{1}{n_{ij} \ P_{i/i,j} \ P_{j/i,j}}$

i.e. weights based on underlined observed proportions. They obtained a goodness of fit statistic of 19.76 on 15 d.f. and were happy to accept the Bradley-Terry Model. However, if the weights are updated using the underlined fitted proportions, the deviance increases to 37 and converges to 32.7 after several iterations.

There is little point in estimating the Π_i's when the model is inadequate. Instead the residuals should be inspected to see if they provide a clue to why the model is not satisfactory.

```
$C FAMILY SIZE CHOICES ,IMREY, JOHNSON AND KOCH , J A S A 1976
$UNITS 42 $DATA X $READ
 2 17
 1 22   0 19
 3 22   1 13   7 11
 1 15 10 10 12 11 13  6
 1 26 11  9 18  6 15  2 17  4
 2 25 13 11 20  6 22  6 14  0 12 11
$FAC K 2 H 21 $CAL K=%GL(2,1):H=%GL(21,2)
:I=H-%GT(H,1)-2*%GT(H,3)-3*%GT(H,6)-4*%GT(H,10)-5*%GT(H,15)
:J=2+%GT(H,1)  +%GT(H,3)  +%GT(H,6)  +%GT(H,10)  +%GT(H,15)
$CAL F=0.5*(%EQ(K,1)-%EQ(K,2))
:L1=F*(%EQ(I,1)-%EQ(J,1)):L2=F*(%EQ(I,2)-%EQ(J,2))
:L3=F*(%EQ(I,3)-%EQ(J,3)):L4=F*(%EQ(I,4)-%EQ(J,4))
:L5=F*(%EQ(I,5)-%EQ(J,5)):L6=F*(%EQ(I,6)-%EQ(J,6))
:L7=F*(%EQ(I,7)-%EQ(J,7))
$YVA X $ERR P $FIT L1+L2+L3+L4+L5+L6+L7+H-%GM $
          SCALED
 CYCLE  DEVIANCE      DF
    4    33.65         15

$C SIGNIFICANT AT 0.5% AS IT EXCEEDS 32.80 , WE CONCLUDE THAT
$C THE BRADLEY TERRY MODEL IS UNACCEPTABLE FOR THIS DATA SET

$C EXAMINE STANDARDISED RESIDUALS
$VAR 21 SR1 SR2 UN
$CAL IN=%GL(%NU,1):INF=IN*%LE(IN,21):INS=%GT(IN,21)*%GL(21,1)
$CAL SR=(%YV-%FV)/%SQRT(%FV) : SR1(INF)=SR : SR2(INS)=SR
: UN(INF)=IN+21 $ACC 2 $ECH $

 UNIT  RESIDUAL  UNIT    RESIDUAL
    1   0.23E-01  22.      0.39
    2  -0.77E-02  23.      0.23
    3   0.34      24.     -0.24
    4  -0.61E-01  25.     -0.23
    5   -2.0      26.      0.45
    6    1.1      27.      0.22
    7    3.1      28.     -0.50
    8  -0.49      29.      0.36
    9  -0.92      30.     -0.61
   10   0.42      31.     -0.64
   11  -0.28      32.      0.23
   12   0.25      33.      0.61E-01
   13   0.43      34.     -0.65E-01
   14  -0.89E-01  35.     -0.20
   15   2.0       36.      0.41
   16   -1.2      37.     -0.33
   17  -0.31      38.      0.77
   18   0.35      39.      1.0
   19   0.27      40.     -1.8
   20  -0.36      41.      0.15E-01
   21   -1.1      42.     -0.16E-01

$C THREE RELATIVELY LARGE RESIDUALS ,AT OBSERVATIONS 7,5 AND 15 .
$RETURN
```

3. EXTENSIONS TO THE MODEL

The basic model may be modified to cope with ties and order effects. Ties occur when each judge is allowed a third option, that of expressing no preference, the probability that he does so being $P_{o/i,j}$. Several authors have proposed models for this situation, including Rao and Kupper (1967) and by Davidson (1970). Whether or not ties occur, it is possible that the order in which items are presented to a judge may influence his or her preference, in which case $P_{i/i,j} \neq P_{j/i,j}$. A model incorporating order effects was suggested by Davidson and Beaver (1977).

Further extension to the multivariate case has been made by Davidson and Bradley (1969) and by Fienberg and Larntz (1976).

The log linear approach can be conveniently applied only to those models which can be expressed multiplicatively, i.e. Davidson for ties, Davidson and Beaver for order and Fienberg and Larntz for the multivariate case.

The Davidson model for ties is multiplicative, and can be expressed in the form:-

$$P_{i/i,j} = c_{ij} (\Pi_i/\Pi_j)^{\frac{1}{2}}$$

$$P_{j/i,j} = c_{ij} (\Pi_j/\Pi_i)^{\frac{1}{2}}$$

$$P_{o/i,j} = c_{ij} \nu$$

where $c_{ij} = 1/((\Pi_i/\Pi_j)^{\frac{1}{2}}+(\Pi_j/\Pi_i)^{\frac{1}{2}}+\nu)$

A loglinear version of this model consists of equations (1) and (2) together with

$$\ln m_{o/i,j} = \mu_{ij} + \ln \nu$$

where now $\mu_{ij} = \ln m + \ln c_{ij}$.

Note that in the limiting case of no ties, $\nu=0$ and $\ln \nu \to -\infty$.

The Fienberg model :-

$$\ln m_{i/i,j} = \mu + \alpha_{ij} + \beta_1$$

$$\ln m_{j/i,j} = \mu + \alpha_{ij} + \beta_2$$

$$\ln m_{o/i,j} = \mu + \alpha_{ij} + \beta_3$$

in which he adopts the traditional rather than the GLIM parametrisation, and insists that

$$\sum \alpha_{ij} = \sum \beta_k = 0$$

has $\beta_3 = -\beta_1-\beta_2$, and the limiting case of no ties corresponds to $-\frac{3}{2}\beta_1 - \frac{3}{2}\beta_2 \to -\infty$.

Indirect estimation of the worth parameters $\hat{\Pi}_i$, and their standard errors is also less straight forward.

The Davidson and Beaver model for items presented in the order i followed by j, and allowing for ties, is

$$P_{i/i,j} = \Pi_i \, b_{ij}$$

$$P_{j/i,j} = \gamma \, \Pi_j \, b_{ij}$$

$$P_{o/i,j} = \nu \, (\Pi_i \Pi_j)^{\frac{1}{2}} \, b_{ij}$$

where $b_{ij} = 1/(\Pi_i + \gamma \Pi_j + \nu (\Pi_i \Pi_j)^{\frac{1}{2}})$

This can be expressed in log linear form as

$$\ln m_{i/i,j} = \mu_{ij} + \tfrac{1}{2} \ln \Pi_i - \tfrac{1}{2} \ln \Pi_j$$

$$\ln m_{j/i,j} = \mu_{ij} - \tfrac{1}{2} \ln \Pi_i + \tfrac{1}{2} \ln \Pi_j + \ln \gamma$$

$$\ln m_{o/i,j} = \mu_{ij} \qquad\qquad\qquad\qquad + \ln \nu$$

where $\mu_{ij} = \ln m + \ln b_{ij} + \tfrac{1}{2} \ln \Pi_i + \tfrac{1}{2} \ln \Pi_j$

Example 3.

Davidson and Beaver give an example of preference testing four food mixes, in which the comparison i followed by j was recorded separately from j followed by i. We have an incomplete 4×4×3 table with diagonal of the 4×4 missing.

TABLE 4

Summary of responses for a preference testing experiment

Items compared in order		no. of comparisons	No. of times i preferred	No. of times j preferred	neither preferred
i	j	n_{ij}	$x_{i/i,j}$	$x_{j/i,j}$	$x_{o/i,j}$
1	2	42	23	11	8
2	1	43	29	6	8
1	3	43	27	11	5
3	1	42	22	14	6
2	3	41	34	6	11
3	2	42	23	16	3
1	4	42	35	6	1
4	1	42	27	11	4
2	4	40	29	9	2
4	2	42	22	15	5
3	4	42	26	11	5
4	3	43	24	14	5

Thus when items 1 and 2 were presented in that order 23 judges preferred 1 to 2 11 preferred 2 to 1, 8 were undecided, but when they were presented in the order 2, 1 29 judges preferred 2 to 1, 6 preferred 1 to 2, 8 were undecided. Evidently there is a tendency for judges to prefer the food mix which was presented first of the pair.

```
$C   FOOD MIXES , DAVIDSON AND BEAVER , BIOMETRICS 1977
$UNITS 48 $DAT X $READ
 0  0  0  23 11  8  27 11  5  35  6  1
29  6  8   0  0  0  34  6  1  29  9  2
22 14  6  23 16  3   0  0  0  26 11  5
27 11  4  22 15  5  24 14  5   0  0  0
$CAL I=%GL(4,12):J=%GL(4,3):K=%GL(3,1):W=%NE(I,J)
$YVA X $ERR P $WEI W
$FAC H 16 $CAL H=%GL(16,3):K1=%EQ(K,1):K2=%EQ(K,2):K3=%EQ(K,3)
$CAL F=0.5*(K1-K2)
:L1=F*(%EQ(I,1)-%EQ(J,1)):L2=F*(%EQ(I,2)-%EQ(J,2))
:L3=F*(%EQ(I,3)-%EQ(J,3)):L4=F*(%EQ(I,4)-%EQ(J,4))
$FIT L1+L2+L3+L4+K2+K3+H-%GM
$C FIT MODEL WITH PARAMETERS FOR WORTHS , ORDER AND TIES
          SCALED
 CYCLE  DEVIANCE       DF
    3    20.94         19

$C THE DAVIDSON-BEAVER MODEL IS ACCEPTABLE
$C NOW CALCULATE ESTIMATES OF THE WORTH PARAMETERS ETC.
$EXTR %PE $CAL G=%EXP(%PE):T=%CU(G):M=G/T(4):%D=%DV:%F=%DF
$DEL %VC %PE $OUT $FIT -L4 $EXTR %VC %PE $OUT 8
$CAL IB=%LE(%GL(231,1),10):IN=%GL(%PL,1)
$CAL VC(IB*%CU(IB))=%VC
$CAL V1=IN-%GT(IN,1)-2*%GT(IN,3)-3*%GT(IN,6)-4*%GT(IN,10)-5*%GT(IN,15)
$CAL V2=1+%GT(IN,1)+%GT(IN,3)+%GT(IN,6)+%GT(IN,10)+%GT(IN,15)
$CAL V12=%EQ(V1,V2)
$MACRO VAR
$CAL %M=M(%I):CVM=V12*%M**2*(M(V2)**2+%EQ(%I,V2))
:CVM=CVM+(1-V12)*2*%M**2*M(V2)*(M(V1)-%EQ(V1,%I))
:VM=CVM*%VC:SM(%I)=%SQRT(%CU(VM)):%I=%I+1:%Q=%LT(%I,3)
$ENDMAC
$CAL %I=1 :%Q=1 $WHILE %Q VAR
$CAL %M=M(%I): VM=VC*%M**2*M(V1)*M(V2)*(2-V12)
:SM(%I)=%SQRT(%CU(VM)):M(5)=%EXP(%PE(4)):M(6)=%EXP(%PE(5))
:SM(5)=M(5)*%SQRT(%VC(10)):SM(6)=M(6)*%SQRT(%VC(15))
$MACRO RED $CAL %R=%DV-%D:%S=%DF-%F
$PRINT %R ' REDUCTION ON ' *-2 %S ' DEGREES OF FREEDOM '$ENDMAC
$MAC TX1 WORTH 1 $END
$MAC TX2 WORTH 2 $END
$MAC TX3 WORTH 3 $END
$MAC TX4 WORTH 4 $END
$MAC TX5 ORDER   $END
$MAC TX6 TIES    $END
$MAC LIN $CAL %M=M(%I):%S=SM(%I)$PRI %1 %M %S $END
$C
$C PARAMETER ESTIMATES AND APPROXIMATE STANDARD ERRORS
 $CAL %I=1$A LIN TX1  LIN $
WORTH 1   0.2623  0.0451
 $CAL %I=2$A LIN TX2  LIN $
WORTH 2   0.3524  0.0734
 $CAL %I=3$A LIN TX3  LIN $
WORTH 3   0.1996  0.0434
 $CAL %I=4$A LIN TX4  LIN $
WORTH 4   0.1858  0.0226
 $CAL %I=5$A LIN TX5  LIN $
ORDER     0.3914  0.0418
 $CAL %I=6$A LIN TX6  LIN $
TIES      0.1657  0.0246
$PRI :$
```

```
$C  FIT MODEL WITHOUT ORDER EFFECT
$FIT -K2 $USE RED$
        SCALED
CYCLE  DEVIANCE      DF
   4    107.4        20

   86.43  REDUCTION ON     1. DEGREES OF FREEDOM
$C FOR ORDER EFFECT    $PRI :

$C  FIT MODEL WITH ORDER AND TIES , WITHOUT PREFERENCES

$FIT K2+K3+H-%GM $USE RED $
        SCALED
CYCLE  DEVIANCE      DF
   4    36.28        22

   15.34  REDUCTION ON     3. DEGREES OF FREEDOM
$C  FOR EQUALITY OF PREFERENCE      $PRI :
```

In the GLIM printout above the data are read in the form of a 4×4×3 table with arbitrary 0 entries for the observations on the diagonal of the 4×4. These observations are ignored in fitting models by being given weight 0, while the others have weight 1.

Note that $\mu_{ij} \neq \mu_{ji}$ in this case. We require H to have a different level for $x_{i/i,j}$ from that for $x_{i/j,i}$.

Otherwise the GLIM statements required are very similar to those of the previous example, with the addition of one column vector in the design matrix which indicates when the terms are presented in inverse order, and another which indicates the ties. The maximum likelihood estimates were obtained in 3 cycles of iteration, and with a deviance of 20.94 on 19 d.f. provide an acceptable fit.

The estimates of the worth, order and tie parameters agree with those found by Davidson and Beaver after 22 iterations of their maximum likelihood program. Standard errors of the parameter estimates cannot be compared since Davidson and Beaver do not report them.

Fitting a model without the order effect gives a deviance of 107.4 on 20 d.f., i.e. a 1 d.f. reduction in deviance of 86.4 is obtained by including the order parameter.

Fitting a model with equal preferences ($\Pi_i = \frac{1}{4}$ or $\lambda_i = 0$) along with one parameter for order and one for ties, gives a deviance of 36.28 on 22 d.f. i.e. a 3 d.f. reduction of 15.34 for preferences.

We conclude that preferences are significantly different and that order is highly significant.

4. CONCLUSION

There is no case to be made for writing an iterative maximum likelihood program to fit multiplicative models to preference data, since GLIM can cope with all known models of this type.

REFERENCES

ATKINSON A.C. (1972). A test of the linear Logistic and Bradley-Terry models. *Biometrika*, 59, 37-42.

BRADLEY, R.A. (1976). Science, Statistics and Paired Comparisons. *Biometrics*, 32, 213-239.

BRADLEY, R.A. and TERRY, M.E. (1952). The rank analysis of incomplete block designs. I. The method of paired comparisons. *Biometrics*, 39, 324-345.

DAVIDSON, R.R. (1970). Extending the Bradley-Terry model to accomodate ties in paired comparison experiments. *JASA*, 65, 317-328.

DAVIDSON, R.R. and Beaver, R.J. (1977). Extending Bradley-Terry model to incorporate within - pair order effects, *Biometrics*, 33, 693-702.

DAVIDSON, R.R. and BRADLEY, R.A. (1969). Multivariate paired comparisons : the extension of a univariate model and associated estimation and test procedures. *Biometrika*, 56, 81-95.

DAVIDSON, R.R. and FARQUHAR, P.H. (1976). A bibliography on the method of paired comparisons. *Biometrics*, 32, 241-252.

FIENBERG, S.E. (1979). Log linear representation for paired comparison models with ties and within-pair order effects. *Biometrics*, 35, 479-481.

FIENBERG, S.E. and LARNTZ, K. (1976) Lof linear representation for paired and multiple comparison models. *Biometrika*, 63, 245-254.

IMREY, P.B., JOHNSON, W.D., and KICH, G.G. (1976). An incomplete contingency-table approach to paired - comparison experiments. *JASA*, 71, 614-623.

NELDER, J.A. and others, (1978). Generalised Interactive Modelling. Release 3 of manual. Numerical Algorithms Group.

RAO, P.V. and KUPPER, L.L. (1967) Ties in paired - comparison experiments : a generalisation of the Bradley-Terry model. *JASA*, 62, 194-204.

ZERMELO, E. (1929) *Math. Zeit.* 29, 436-460.

A GLM FOR Estimating Probabilities in Retrospective Case-Control Studies

M. SLATER   and   R. D. WIGGINS
Computer Science   Social Science and
and Statistics,   Business Studies
Queen Mary College,   Polytechnic of Central London,
(University of London),   London w1
Mile End Road,
London E1

SUMMARY
A simple GLM for estimating probabilities of disease incidence
across subgroups of a population from retrospective case-control
studies is proposed. The method is illustrated by an application
of GLIM to a cross classification table from a study of cot
deaths in Lambeth.

Keywords: RETROSPECTIVE STUDIES; CASE-CONTROL; GLIM; GLM

1. INTRODUCTION

The retrospective study is an appropriate research method when
the proportion of 'cases' in the population under study is small.
Prospective studies in such circumstances would be prohibitively
expensive both financially and in time.

One objective of a retrospective study of, for example, disease
incidence in a population, would be to identify the combination
of factors which contribute to individuals' enhanced risk of
disease. Wiggins and Slater (1981) examined an estimator based on
retrospective data for the risk across different subgroups of a
population, using an initial approach described by Cornfield
(1956) and MacMahon and Pugh (1970, p.274). Risk was defined as
the probability of an individual in a subgroup becoming a case.

The present paper reexamines the problem in the context of
generalised linear models (Nelder and Wedderburn, 1972). It is
shown that a surprisingly simple GLM is appropriate for the
analysis of retrospective case-control data, of which the Wiggins
and Slater estimator turns out to be a special case. The method
is illustrated with data from a study of sudden infant deaths
(cot deaths) in Lambeth carried out by the Department of
Community Medicine, St Thomas's Hospital, London (Palmer **et al**,
1980).

2. METHOD

Let R be the total number of cases of the disease in the
population over the time period of study. Typically n controls
are selected at random from the remainder of the population.
Suppose that the population is partitioned into k exclusive
subgroups corresponding to a cross-classification reflecting
factors thought to be of importance in the aeteology of the
disease. The problem is to estimate p_i, the probability of an
individual in the ith subgroup becoming a case. In a
retrospective study r_i will be all of the cases in the ith
subgroup during the time period of the study, and n_i the
corresponding number of controls. Let N_i be the actual number of

non cases in the ith subgroup in the population as a whole. Then clearly

$$p_i = \frac{r_i}{r_i + N_i} \quad i = 1, 2, \ldots, k \qquad (2.1)$$

Given the usually small sampling fraction adopted, the n_i will be very much smaller than the true N_i. N_i may be estimated by

$$\hat{N_i} = \frac{n_i}{n} R \frac{(1 - \epsilon)}{\epsilon} \qquad (2.2)$$

where ϵ is the known proportion of cases in the population. Substituting (2.2) into (2.1) leads to the estimator

$$\hat{p_i} = \frac{n r_i \epsilon}{n r_i \epsilon + R n_i (1 - \epsilon)} \qquad (2.3)$$

Wiggins and Slater (op. cit) found approximations for the expectation and bias of (2.3) conditional on the r_i and ϵ.
Let λ_i be the probability that an individual who is a case is in the ith cell, and π_i be the corresponding probability for a control. That is,

$$\lambda_i = \Pr(i \mid \text{case})$$
$$\pi_i = \Pr(i \mid \text{control})$$

Hence,

$$p_i = \frac{\lambda_i \epsilon}{\lambda_i \epsilon + \pi_i (1 - \epsilon)}$$

$$\therefore \quad \pi_i = \lambda_i \zeta \frac{(1 - p_i)}{p_i}$$

$$\zeta = \epsilon / (1 - \epsilon)$$

Now suppose that p_i varies with a set of explanatory variables $x_1, x_2, \ldots, x_q$, according to the linear predictor

$$\eta_i = \sum_{j=1}^{q} \beta_j x_{ij}$$

and such that

$$p_i = \frac{1}{1 + e^{\eta_i}} \qquad (2.4)$$

Then,

$$\frac{1 - p_i}{p_i} = e^{\eta_i}$$

so that,

$$\pi_i = \lambda_i e^{\eta_i}$$

Let μ_i be the means of the observed n_i , giving the relationship,

$$E(n_i) = \mu_i$$
$$= n \pi_i$$
$$= n \lambda_i \zeta e^{\eta_i}$$

The above defines a GLM with

1. A Poisson error structure on the n_i ;

2. the linear predictor η_i ;

3. the inverse link function

$$\mu_i = n \lambda_i e^{\eta_i} \zeta \qquad (2.5)$$

The link function is clearly

$$\eta_i = \ln \left(\frac{\mu_i}{n \lambda_i \zeta} \right) \qquad (2.6)$$

with derivative

$$\frac{d\eta}{d\mu} = \frac{1}{\mu} \qquad (2.7)$$

It is also possible, of course to find estimates of the p_i from estimates of the linear predictor η_i, which are obtainable from the GLIM analysis, by computing p_i as given in (2.4).

3. THE GLIM APPROACH

The GLM defined in section 2 is not a standard GLIM model since the link function is not one of those provided explicitly by GLIM. However, GLIM does provide users with the capability of constructing their own combination of probability distribution and link function (Baker and Nelder, 1978, section 18). This may be done by specifying four MACROs, and then using the $FIT directive in the usual way.

The four MACROs required for the GLM given above are as follows:
1. The system vector %FV , must be assigned the fitted values calculated from the linear predictor %LP. Using %N as the scalar identifier holding the value of n, and L the vector of the λ_i , then from (2.5)

$$\%FV = \%N*L*\%Q*\%EXP(\%LP)$$

where $\%Q$ is the scalar identifier for $\bar{\lambda}$.
2. The system vector $\%DR$ must be assigned the values of the derivatives of the link function, so that from (2.7)

$$\%DR = 1/\%FV$$

3. MACROs 3 and 4 will be exactly the same as those given in the GLIM manual, since they must provide the variance function and deviance of the Poisson distribution. (See Baker and Nelder, op cit. section 18.2).

Two further vectors need to be set: first, the linear predictor must be initialized. Let NI be the vector holding the values of the n_i, then from (2.6)

$$\%LP = \%LOG(NI/(NI*\%Q))$$

Finally, values must be assigned to the vector L. Over the time period of the study the relative frequencies r_i/R are estimates of λ_i. However, these would not be the only or best estimators that could be used. In fact for the Lambeth cot death data used in this paper for illustrative purposes, the relative frequencies lead to an unacceptable estimate (zero) of the probability of cot death in one of the subgroups, whereas obviously it is not impossible for a cot death to occur in this subgroup (see Table 1).

To estimate the probabilities a method discussed by Good (1965) is used. Suppose initially it was judged that a case is equally likely to be in any of the k subgroups. Then having observed the frequencies, the updated subgroup probability estimates would be

$$\frac{r_i + \phi}{R + k\phi}$$

where the size of ϕ determines the variance of the estimates. The larger the value of ϕ, the greater the weight attached to the initial 'equally likely' estimates. ϕ may be estimated by choosing that value which minimises the deviance of the model under consideration.

4. ILLUSTRATIVE EXAMPLE

Table 1 reproduces some data from the Lambeth study of cot deaths. This was a retrospective case control study, where over a two year period all of the cot deaths in the Lambeth area were identified (R = 54), and a random sample of (n = 55) controls selected. The table shows the number of cases and controls for four maternal age groups, with one or no previous pregnancies. For the Lambeth area over this time period ϵ was known to be 0.0059. In this example k = 8, but the method can obviously be extended to cover more cells or factors.

The GLIM MACROs and output are given in Appendix A. From an initial analysis plotting the deviance of the full first order model (i.e., ROWS+COLS) against different values of ϕ, ϕ was estimated to be 0.50 (MACRO GETD). Subsequent fits revealed a significant decrease in deviance from the null model to the model including only the row factor (age); adding the column factor (parity) was marginally significant.

Using the saturated model (i.e., ROW + COL + ROW.COL) the probability estimates obtained are the same as those obtained by Wiggins and Slater (allowing for the adjustment in different r_i/R

used). However, what this paper demonstrates is that we can now obtain a parsimonious model of the data, from which probability estimates can be produced. The earlier estimation technique is equivalent to a model which exactly reproduces the data. For the Lambeth data the first order model gives an adequate fit and the probability estimates obtained can be interpreted as the product of independent row and column effects. These probability estimates are given in Table 2.

It is worth comparing these results with those that can be obtained using an alternative approximate method. Suppose the true numbers of controls (N_i) were known for each cell. Then treating r_i as the dependent variable, the usual binomial logistic model would be appropriate, with $N_i + r_i$ as the binomial denominator. The N_i may be estimated using (2.2). The results of this analysis are given in Appendix B. It is notable that they reproduce almost exactly those obtained by the previous analysis.

REFERENCES

BAKER, R.J. and NELDER, J.A.(1978) The GLIM System, Release 3. Oxford: Numerical Algorithms Group.

CORNFIELD, J.(1956) A Statistical Problem Arising from Retrospective Studies, Proceedings of the Third Berkeley Symposium on Mathematical Statistics and Probability, Vol. 4 J. Neyman (ed) pp 135-148. University of California Press.

GOOD, I.J.(1965) The Estimation of Probabilities ,Cambridge,MIT

NELDER J.A. and WEDDERBURN, R.W.M.(1972) Generalized Linear Models. J.R. Statit. Soc. B., 42, 109-142.

PALMER, S.R., WIGGINS, R.D., and BEWLEY, B.R.(1980) Infant Deaths in Inner London: a health Care Planning Team Study, Community Medicine 2.

WIGGINS, R.L. and SLATER, M.(1981) Estimating Probabilities from Retrospective Data with an Application to Cot Death in Lambeth. Biometrics 37, 2, pp 377-382.

TABLE 1

Lambeth 'cot deaths' and controls
maternal age by previous pregnancies

| maternal age (years) | previous pregnancies | | | | totals | |
| | 0 | | 1+ | | | |
	deaths	controls	deaths	controls	deaths	controls
<20	7	5	6	4	13	9
20-24	4	8	17	11	21	19
25-29	0	2	4	11	4	13
total	16	21	38	34	54	55

TABLE 2

Estimated Probabilities of 'Cot Death'
by Maternal Age and previous pregnancies

Probabilities shown as risk per 1000

| maternal age | previous pregnancies | |
	0	1+
<20	6.8 (8.4)	12.6 (9.0)
20-24	4.1 (3.0)	7.6 (9.3)
25-29	4.7 (5.0)	8.7 (8.2)
>30	1.3 (0)	2.3 (2.1)

The Wiggins and Slater estimates are shown in brackets

```
APPENDIX A
----------
$SUBFILE OWN
$C SET UP THE DATA
    R IS VECTOR OF CASE FREQUENCIES; NI ARE THE CONTROLS
    %E IS EPSILON
$UNITS 8 $FACTOR ROW 4 COL 2 $DATA R NI $READ
7 5 6 4 4 8 17 11
5 6 11 8 0 2 4 11

$CAL ROW = %GL(4,2) : COL = %GL(2,1)
  :  %E = 0.0059   :   %Q = %E/(1-%E)

$YVAR NI $

$C CALCULATE THE LAMBDA (L) WHERE %Z IS PHI
$MACRO LCAL $CAL L = (R+%Z)/(54+8*%Z) $ $ENDMACRO


$MACRO M1 $CAL %FV = 55*L*%EXP(%LP)*%Q $ $END
$MACRO M2 $CAL %DR = 1/%FV $ $END
$MACRO M3 $CAL %VA = %FV $ $END
$MACRO M4 $CAL %DI = 2*(%YV*%LOG(%YV/%FV)-(%YV-%FV)) $ $END


$MACRO INIT
$C COMPUTE THE LINEAR PREDICTOR
  $CAL %A = 55*%Q : %LP = %LOG(NI/(%A*L)) $
$END

$C COMPUTE THE ESTIMATED PROBABILITIES
$MACRO PROB $CAL P = 1/(1+%EXP(%LP)) $LOOK P $ $ENDMACRO

$OWN M1 M2 M3 M4 $

$VAR 20 PHI DEV
$MACRO GETD
$C TO COMPUTE DEVIANCES FOR DIFFERENT VALUES OF
PHI. %K MUST BE INITIALIZED AT 20
  $CAL %I = 21-%K : %Z = PHI(%I) $USE LCAL $USE INIT
  $FIT ROW+COL
  $CAL DEV(%I) = %DV
  :   %K = %K-1
  $
$END

$MACRO GO $USE LCAL $USE INIT $FIT $USE PROB $FIT ROW+COL $USE PROB$ $END
$RETURN
$FINISH
```

```
$CAL PHI = %CU(1)/20 + 0.1
  !    %K = 20
$OUTPUT 6 $ECHO $WHILE %K GETD $
PLOT OF DEVIANCES AGAINST PHI
```

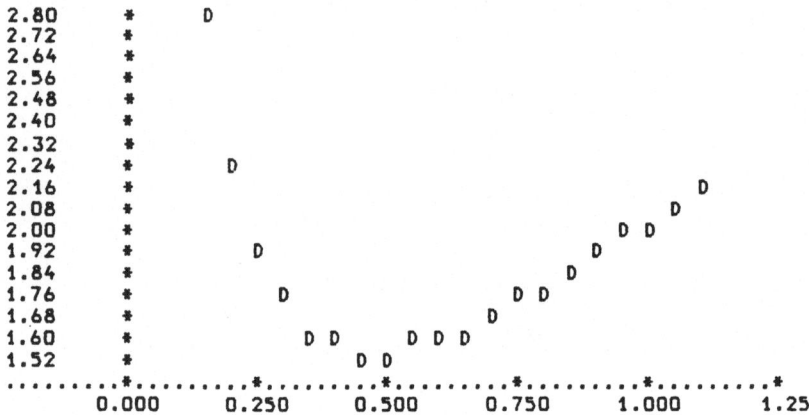

```
VALUES OF PHI AND CORRESPONDING DEVIANCES
  1    0.1500        2.802
  2    0.2000        2.268
  3    0.2500        1.948
  4    0.3000        1.754
  5    0.3500        1.638
  6    0.4000        1.575
  7    0.4500        1.549
  8    0.5000        1.548
  9    0.5500        1.565
 10    0.6000        1.595
 11    0.6500        1.635
 12    0.7000        1.682
 13    0.7500        1.734
 14    0.8000        1.790
 15    0.8500        1.848
 16    0.9000        1.907
 17    0.9500        1.967
 18    1.0000        2.028
 19    1.050         2.088
 20    1.100         2.148
```

```
$C PHI OF ABOUT 0.5 GIVES MIN DEVIANCE
    FIT MODELS WITH THIS VALUE


$CAL %Z = 0.5
$C THE FOLLOWING MACRO PRODUCES FIRST THE NULL FIT AND THEN FULL FIT
    $USE GO$
----- CURRENT DISPLAY INHIBITED
 CYCLE  DEVIANCE       DF
    3    17.34         7

    1   0.5900E-02
    2   0.5900E-02
    3   0.5900E-02
    4   0.5900E-02
    5   0.5900E-02
    6   0.5900E-02
    7   0.5900E-02
    8   0.5900E-02
 CYCLE  DEVIANCE       DF
    4    1.548         3

    1   0.6848E-02
    2   0.1255E-01
    3   0.4128E-02
    4   0.7583E-02
    5   0.4697E-02
    6   0.8623E-02
    7   0.1271E-02
    8   0.2341E-02
 $FIT COL$
 CYCLE  DEVIANCE       DF
    3    16.08         6

 $FIT ROW$
 CYCLE  DEVIANCE       DF
    4    5.616         4

 $FIT ROW+COL $USE PROB$
 CYCLE  DEVIANCE       DF
    3    1.548         3

    1   0.6848E-02
    2   0.1255E-01
    3   0.4128E-02
    4   0.7583E-02
    5   0.4697E-02
    6   0.8623E-02
    7   0.1272E-02
    8   0.2342E-02
 $FIT ROW*COL $USE PROB$
 CYCLE  DEVIANCE       DF
    3   0.3624E-04     0

    1   0.8371E-02
    2   0.9063E-02
    3   0.3156E-02
    4   0.8874E-02
    5   0.5133E-02
    6   0.8025E-02
    7   0.1405E-02
    8   0.2297E-02
$STOP
```

APPENDIX B

```
$C NOW FIT BINOMIAL MODEL
    FIRST ESTIMATE TOTAL NUMBER OF INDIVIDUALS IN EACH CELL
    ESTIMATED (POPULATION) NUMBER OF CONTROLS + NUMBER OF CASES

$CAL NEWN = (54/%E)*(NI/55) + R $

$C DECLARE BINOMIAL ERROR STRUCTURE WITH DEFAULT LINK

$YVAR R $ERROR B NEWN

$C FIT SAME MODELS AS IN PREVIOUS CASE

$FIT $CAL NEWP = %FV/NEWN $PRINT NEWP $
         SCALED
  CYCLE  DEVIANCE       DF
    4     17.82          7

  0.0059  0.0059  0.0059  0.0059  0.0059  0.0059  0.0059  0.0059

$FIT COL$
         SCALED
  CYCLE  DEVIANCE       DF
    4     16.09          6

$FIT ROW$
         SCALED
  CYCLE  DEVIANCE       DF
    4     7.276          4

$FIT ROW+COL $CAL NEWP = %FV/NEWN $PRINT NEWP$
         SCALED
  CYCLE  DEVIANCE       DF
    3     2.688          3

  0.0062  0.0116  0.0044  0.0082  0.0046  0.0085  0.0011  0.0020

$FIT ROW*COL $CAL NEWP = %FV/NEWN $PRINT NEWP$
         SCALED
  CYCLE  DEVIANCE       DF
   10    0.4523E-04      0

  0.0083  0.0089  0.0030  0.0092  0.0050  0.0082  0.0000  0.0022
$STOP
```

Lecture Notes in Statistics